D0744730

Practical Data Science with Python 3

Synthesizing Actionable Insights from Data

Ervin Varga

Apress®

Practical Data Science with Python 3: Synthesizing Actionable Insights from Data

Ervin Varga
Kikinda, Serbia

ISBN-13 (pbk): 978-1-4842-4858-4   ISBN-13 (electronic): 978-1-4842-4859-1
https://doi.org/10.1007/978-1-4842-4859-1

Copyright © 2019 by Ervin Varga

Managing Director, Apress Media LLC: Welmoed Spahr
Acquisitions Editor: Celestin Suresh John
Development Editor: James Markham
Coordinating Editor: Aditee Mirashi

Cover designed by eStudioCalamar

Cover image designed by Freepik (www.freepik.com)

Distributed to the book trade worldwide by Springer Science+Business Media New York, 233 Spring Street, 6th Floor, New York, NY 10013. Phone 1-800-SPRINGER, fax (201) 348-4505, e-mail orders-ny@springer-sbm.com, or visit www.springeronline.com. Apress Media, LLC is a California LLC and the sole member (owner) is Springer Science + Business Media Finance Inc (SSBM Finance Inc). SSBM Finance Inc is a **Delaware** corporation.

For information on translations, please e-mail rights@apress.com, or visit http://www.apress.com/rights-permissions.

Apress titles may be purchased in bulk for academic, corporate, or promotional use. eBook versions and licenses are also available for most titles. For more information, reference our Print and eBook Bulk Sales web page at http://www.apress.com/bulk-sales.

Any source code or other supplementary material referenced by the author in this book is available to readers on GitHub via the book's product page, located at www.apress.com/978-1-4842-4858-4. For more detailed information, please visit http://www.apress.com/source-code.

Printed on acid-free paper

Like traveling, writing a book is more enjoyable when accompanied by your family. I am thankful to my wife, Zorica, and sons, Andrej and Stefan, for all their great support.

Table of Contents

About the Author xi

About the Technical Reviewer xiii

Acknowledgments xv

Introduction xvii

Chapter 1: Introduction to Data Science.....................................1

Main Phases of a Data Science Project 2

Brown Cow Model Case Study 4

Big Data 9

Big Data Example: MOOC Platforms 10

How to Learn Data Science..12

Domain Knowledge Attainment—Example 14

Programming Skills Attainment—Example 16

Overview of the Anaconda Ecosystem...18

Managing Packages and Environments 20

Sharing and Reproducing Environments 23

Summary..25

References...26

Chapter 2: Data Engineering..29

E-Commerce Customer Segmentation: Case Study 30

Creating a Project in Spyder..33

Downloading the Dataset 34

Exploring the Dataset 36

Inspecting Results 56

Persisting Results 60

Restructuring Code to Cope with Large CSV Files 62

Public Data Sources 64

Summary 69

References 70

Chapter 3: Software Engineering 73

Characteristics of a Large-Scale Software System 75

Software Engineering Knowledge Areas 79

Rules, Principles, Conventions, and Standards 81

Context Awareness and Communicative Abilities 85

Reducing Cyclomatic Complexity 89

Cone of Uncertainty and Having Time to Ask 91

Fixing a Bug and Knowing How to Ask 93

Handling Legacy Code 102

Understanding Bug-Free Code 103

Understanding Faulty Code 105

The Importance of APIs 107

Fervent Flexibility Hurts Your API 109

The Socio-* Pieces of Software Production 111

Funny Elevator Case Study 112

Summary 118

References 119

Chapter 4: Documenting Your Work...**121**

JupyterLab in Action 125

    Experimenting with Code Execution.. 126

    Managing the Kernel 134

    Descending Ball Project.. 136

    Refactoring the Simulator's Notebook....................................... 148

Document Structure.. 150

    Wikipedia Edits Project.. 152

Summary.. 156

References 157

Chapter 5: Data Processing **159**

Augmented Descending Ball Project... 159

    Version 1.1.. 160

    Version 1.2.. 172

    Version 1.3.. 184

Abstractions vs. Latent Features 200

    Compressing the Ratings Matrix 201

Summary.. 205

References 207

Chapter 6: Data Visualization **209**

Visualizing Temperature Data Case Study................................... 210

    Showing Stations on a Map... 211

    Plotting Temperatures 213

Closest Pair Case Study 218

    Version 1.0.. 223

    Version 2.0.. 229

    Version 3.0.. 233

Enquiry of Algorithms Evolution 241

Interactive Information Radiators 241

The Power of Domain-Specific Languages 244

Summary 252

References 253

Chapter 7: Machine Learning 255

Exposition of Core Concepts and Techniques 258

Overfitting 271

Underfitting and Feature Interaction 276

Collinearity 278

Residuals Plot 281

Regularization 285

Predicting Financial Movements Case Study 286

Data Retrieval 288

Data Preprocessing 288

Feature Engineering 303

Implementing Streaming Linear Regression 308

Summary 314

References 316

Chapter 8: Recommender Systems 317

Introduction to Recommender Systems 318

Simple Movie Recommender Case Study 322

Introduction to LensKit for Python 329

Summary 338

References 339

Chapter 9: Data Security 341

Checking for Compromise 342

Introduction to the GDPR 349

Machine Learning and Security 359

    Membership Inference Attack 359

    Poisoning Attack 363

Summary 365

References 366

Chapter 10: Graph Analysis 369

Usage Matrix As a Graph Problem 370

    Opposite Quality Attributes 376

    Partitioning the Model into a Bipartite Graph 377

    Scalable Graph Loading 380

Social Networks 385

Summary 395

References 396

Chapter 11: Complexity and Heuristics 397

From Simple to Complicated 401

    Counting the Occurrences of a Digit 402

    Estimating the Edge Betweenness Centrality 409

    The Count of Divisible Numbers 413

From Disorder to Complex 415

    Exploring the KDD Cup 1999 Data 416

Cynefin and Data Science 420

Summary 425

References 425

Chapter 12: Deep Learning **427**

Intelligent Machines...428

Intelligence As Mastery of Symbols 431

    Manual Feature Engineering 432

    Machine-Based Feature Engineering 436

Summary..449

References 449

Index..**451**

About the Author

Ervin Varga is a Senior Member of IEEE and Professional Member of ACM. He is an IEEE Software Engineering Certified Instructor. Ervin is an owner of the software consulting company Expro I.T. Consulting, Serbia. He has an MSc in computer science, and a PhD in electrical engineering (his thesis was an application of software engineering and computer science in the domain of electrical power systems). Ervin is also a technical advisor of the open-source project Mainflux.

About the Technical Reviewer

 **Jojo Moolayil** is an artificial intelligence professional and published author of three books on machine learning, deep learning, and IoT. He is currently working with Amazon Web Services as a Research Scientist – AI in AWS's office in Vancouver, BC.

Jojo was born and raised in Pune, India, and graduated from the University of Pune with a major in Information Technology Engineering. His passion for problem-solving and data-driven decision-making led him to start a career with Mu Sigma Inc., the world's largest pure-play analytics provider. There, he was responsible for developing machine learning and decision science solutions for large, complex problems for healthcare and telecom giants. He later worked with Flutura (an IoT analytics startup) and General Electric with a focus on industrial AI, in Bangalore, India.

In his current role with AWS, he works on researching and developing large-scale AI solutions for combating fraud and enriching the customer's payment experience in the cloud. He is also actively involved as a tech reviewer and AI consultant with leading publishers and has reviewed over a dozen books on machine learning, deep learning, and business analytics.

You can reach out to Jojo at https://www.jojomoolayil.com/, https://www.linkedin.com/in/jojo62000, and https://twitter.com/jojo62000.

Acknowledgments

I would like to thank Apress for giving me an opportunity and full support for writing this book about data science. Comments and help from James Markham, Aditee Mirashi, and Celestin Suresh John were invaluable.

I am also grateful for excellent remarks from Jojo John Moolayil, who was the technical reviewer on this book.

Introduction

This book amalgamates data science and software engineering in a pragmatic manner. It guides the reader through topics from these worlds and exemplifies concepts through software. As a reader, you will gain insight into areas rarely covered in textbooks, since they are hard to explain and illustrate. You will see the Cynefin framework in action via examples that give an overarching context and systematic approach for your data science endeavors.

The book also introduces you to the most useful Python 3 data science frameworks and tools: Numpy, Pandas, scikit-learn, matplotlib, Seaborn, Dask, Apache Spark, PyTorch, and other auxiliary frameworks. All examples are self-contained and allow you to reproduce every piece of content from the book, including graphs. The exercises at the end of each chapter advise you how to further deepen your knowledge.

Finally, the book explains, again using lots of examples, all phases of a data science life cycle model: from project initiation to data exploration and retrospection. The aim is to equip you with necessary comprehension pertaining to major areas of data science so that you may *see the forest for the trees*.

CHAPTER 1

Introduction to Data Science

Let me start by making an analogy between software engineering and data science. Software engineering may be summarized as the application of engineering principles and methods to the development of software. The aim is to produce a dependable software product. In a similar vein, data science may be described as the application of scientific principles and methods to data collection, analysis, and reporting. The goal is to synthesize reliable and actionable insights from data (sometimes referred as *data product*). To continue with our analogy, the *systems/software development life cycle (SDLC)* prescribes the major phases of a software development process: project initiation, requirements engineering, design, construction, testing, deployment, and maintenance. The *data science process* also encompasses multiple phases: project initiation, data acquisition, data preparation, data analysis, reporting, and execution of actions (another "phase" is data exploration, which is more of an all-embracing activity than a stand-alone phase). As in software development, these phases are quite interwoven, and the process is inherently iterative and incremental. An overarching activity that is indispensable in both software engineering and data science (and any other iterative and incremental endeavor) is *retrospection*, which involves reviewing a project or process to determine what was successful and what could be improved. Another similarity to software engineering is that data science also relies

© Ervin Varga 2019
E. Varga, *Practical Data Science with Python 3*,
https://doi.org/10.1007/978-1-4842-4859-1_1

on a multidimensional team or team of teams. A typical project requires domain experts, software engineers specializing in various technologies, and mathematicians (a single person may take different roles at various times). Yet another common denominator with software engineering is a penchant for automation (via programmability of most activities) to increase productivity, reproducibility, and quality. The aim of this chapter is to explain the key concepts regarding data science and put them into proper context.

Main Phases of a Data Science Project

Figure 1-1 illuminates the major phases of a data science process. These phases shouldn't be treated in a waterfall fashion (meaning one phase must be completed before the next begins). Rather, a typical data science project involves many iterations, similarly to an Agile software development project. The concept of phases is intended only to remind us that focus shifts over time from one phase to another. It isn't evident from looking at Figure 1-1, but the most critical stage is project initiation. To quote John Dewey (*The Theory of Inquiry*, 1938):

> *It is a familiar and significant saying that a problem well put is half-solved.*

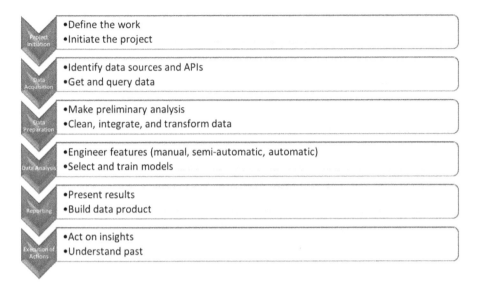

Figure 1-1. *The main phases of a data science project*

The *Brown Cow* model from the Volere requirements engineering method is a helpful resource to explain the project initiation phase (see reference [1] in the "References" section at the end of the chapter). Figure 1-2 represents an adaptation of this model that depicts what happens when a data science project is initiated. The mere fact that we have some data to play with doesn't automatically warrant a full-blown project. It is important to properly define the problem, assess the situation (risks, assets, resources, contingencies), and describe the goals, including evaluation criteria. We must end up with well-formulated research questions and have a good grasp of implementation details. The whole endeavor must be fathomed in a holistic fashion. The beginning must be connected to the end, and the end must instigate further inquiries.

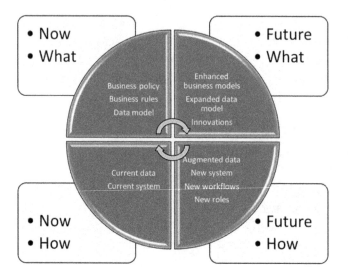

Figure 1-2. *The Brown Cow model adapted for a data science project*

The Brown Cow model contains four segments: the *How* tackles the solution space, the *What* touches upon the problem space, the *Now* designates the current situation, while the *Future* describes the desired state. Each segment designates one specific viewpoint, which helps to avoid confusion among stakeholders. The Brown Cow model incorporates a systematic procedure for transitioning from the current state toward the future, while satisfying the interests of key stakeholders of a project. Without their global consensus, it is very hard to judge the successfulness of the final data product. The next section illustrates this model in a case study.

Brown Cow Model Case Study

You may be wondering whether data science is just a new fad on the horizon. If you examine this question in the light of the previous section's first paragraph, then you might conclude that it isn't new at all. A superb application of data science was done by Dr. John Snow, a father of modern epidemiology. His study of interest was related to a cholera outbreak in London in 1854 (see [2] in the "References" section for more details). He followed the data science process entirely, as described here (the numbers 1 and 2 denote increments that resulted in major milestones):

> **Project initiation**: Based on symptoms of the disease (vomiting and diarrhea), Dr. Snow properly surmised that it must be caused by an infection carried by something people ate or drank. Prevalent public opinion at that time was that cholera was transmitted by bad smells (a.k.a. miasma theory). According to the Brown Cow model, the Now-How is the existing registry of deaths as well as total confusion regarding what is the root cause of cholera. The Now-What is the prevalent wrong theory of how the disease spreads (with a futile rule to cover your nose and mouth). The Future-What is the new theory of what may cause cholera. The Future-How is the incremental procedure to establish causation between contaminated water and cholera as well as clear advice about what should be the immediate next steps. It is also important to note that this revelation opened up the possibility to further investigate cholera, which eventually led to the discovery of the matching bacteria.

> **Data engineering 1 (acquisition, preparation, and exploration)**: Snow recorded the location of each death around the Soho district of London (where the epidemy had emanated) and put everything on a map (as shown in Figure 1-3). He also marked the water pumps on a map to be able to discern

patterns. This is a fine example of using visualization in parallel to raw tabular data; exploration most frequently entails some sort of graphics (image, plot, graph, etc.). Exploration as an activity permeates all phases of the data science process, though.

Figure 1-3. *Snow's original map. Black bars denote deaths. Black discs are pumps. The clustering of death cases around the Broad Street pump is apparent.*

Data analysis 1: After Snow carefully examined every case (including important outliers), he established a strong correlation between deaths and the Broad Street pump. He was cautious and resisted establishing causation at this stage. This is a fine example of applying a scientific skillset in working with data (the Web is full of embarrassing stories of correlation being misapplied as causation).

Retrospective: The first milestone was an enabler for the second part.[1] Snow had revisited the original plan and prepared the next stage to answer his initial research question. He needed to carry out a *randomized control experiment* (the best way to assuredly reach causality) and had to use the method of *comparison*. In a randomized experiment, participants are segregated into two groups: treatment (e.g., those who drank contaminated water) and control (e.g., those who were not exposed to infection). The comparison method seeks to find an association between the applied treatment and observed outcome. It is very important for groups to systematically differ only in that single characteristic that is the criteria for separation. In Snow's case, this was the water supply received by people in these categories. Luckily, there were two water suppliers whose customers weren't

---

[1]A retrospective gives you and your team an opportunity to take notes, make comparisons, hold meetings, and overall improve the current process. In Agile projects, each iteration closes with a retrospective, where the team contemplates what worked well and what went wrong and makes pragmatic steps to enhance the current way of working. A data science team may use a similar strategy to smooth out process-related difficulties.

one-sidedly different in any other aspect except water supply. One supplier delivered clean water upriver from the sewage discharge point, and the other drew its contaminated water below it. The groups were naturally randomized and formed.

Data engineering 2: Snow collected data on all cases of cholera for a broader area of London that were covered by these water suppliers.

Data analysis 2: After a careful analysis, it was evident that people got sick by drinking contaminated water. This was the moment when Snow could safely claim a causal relationship between an infection through contaminated water and cholera.

Reporting: Snow prepared a detailed table showing death rates of people belonging to the two groups. The death rate in the treatment group was ten times higher than that of the control group, so he was confident in his statement.

Action: The authorities removed the handle from the Broad Street pump to prevent further infections, and it proved effective. Further investigations and actions followed this study.

Evidently, this is a compelling data science project from the 19th century. Why, then, is data science such a hot topic nowadays? The answer lurks in the name of the discipline itself (*data* science). It is popular again due to the vastly different data and concomitant complex problem space(s) embodied by *Big Data* (see also the sidebar "Big Data Requires Data Scientists"). In the past, data was relatively scarce and data management solutions were much more expensive. Today, we have a *data deluge* phenomenon that was aptly commented on by John Naisbitt as follows: "We are drowning in information and starving for knowledge."

BIG DATA REQUIRES DATA SCIENTISTS

Big Data, covered in the next section, refers to data at a scale that is difficult to conceptualize. An analogy to help you understand why data scientists are required when designing a Big Data system is that of designing a building. If you want to design your own home, you might be able to draft your own floor plan (or find one on the Internet) and give it to the builder; you don't need to be an architect. Scale that up to designing a multistory apartment building, then you must be a certified architect, but you can rely on existing resources and tools to complete the design.

Finally, if you're hired to design the world's tallest skyscraper or build apartments on an artificial island in the middle of a sea, then you not only must be a certified architect, but also must possess extraordinary knowledge and experience, apply unconventional methods, and devise unique tools. Designing a Big Data system is analogous to designing a skyscraper and requires data scientists with vast knowledge and experience.

Big Data

The notion of being *big* has multiple dimensions. In the realm of Big Data this is articulated as four Vs:

- *Volume* denotes the sheer amount of data. The assumption is that the vast quantity of data cannot fit on a single machine (not even on disk, let alone in memory), so the data must be distributed over dozens of networked machines. This brings in all sorts of issues related to distributed computing that don't exist in the case of a single machine.

- *Variety* designates the property of data being organized in various ways. In a classical setup, we presume well-structured data, whose schema is properly documented.

Usually such data resides in relational databases. With Big Data, we must also deal with unstructured and semi-structured forms. Nonetheless, all data eventually needs to be aligned and managed in a unified fashion.

- *Velocity* dictates the pace of data changes (arrival of new data, update of existing data, and removal of data). Besides processing data at rest, many times we must handle data changes in real time to avoid any data loss. This kind of data management is known as *stream processing*. Streams of data may arrive to a system previously trained on historical data and this data combination (historical + real-time) is sometimes called *actionable information*.

- *Veracity* is about trustworthiness of data. As we amalgamate disparate data sources, we must handle inaccurate, incomplete, and misapplied (sometimes adversary) data.

All these Vs require novel methods, algorithms, and technologies. Also, the complexity and size of the required software systems become larger. These are the principal reasons why data science gets so much attention from both research communities and industry. The next example illustrates these dimensions.

Big Data Example: MOOC Platforms

A massive open online course (MOOC) is an online course offered via a MOOC platform to a large, worldwide community. Some of the most popular platforms are Coursera, edX, Udacity, Khan Academy, and Stanford Online. (Most courses are free, but some require payment.) Generating actionable insights on top of a MOOC platform belongs to the

Big Data problem space. Table 1-1 shows how Dr. Snow's cholera project differs from a data science project built on top of a MOOC data-generation platform in regard to the four Vs of Big Data.

Table 1-1. *Differences Between Old and Modern Data Science Projects in Light of Big Data*

	Snow's Project	MOOC Platform–Based Project
Volume	The total impacted population was of a manageable size to be tracked by a single person.	One platform may host hundreds of courses, each potentially having thousands of students. A course is composed of multimedia material as well as content created by participants (comments on the associated discussion forum, exam answers, assignment submissions, etc.). Just to store the data, you need a huge distributed file system.
Variety	The data was well defined as a set of locations of deaths caused by cholera.	All sorts of data are present: structured course material, unstructured discussion forum stuff, and semi-structured assignment submissions (just to name a few).
Velocity	The rate of death cases was low enough to be tracked by a person (including visualization effort).	The system experiences an extraordinary amount of activity by its users. Tracking their actions, serving course content, and evaluating exams/assignments require a powerful cloud infrastructure.
Veracity	The data was absolutely reliable.	On a discussion forum, you can find all kinds of comments. To bring order to that mess, students upvote or downvote posts and course staff also make comments (sometimes simply to endorse a student's opinion). There are also guidelines about how to name new threads, how to avoid duplication, how to behave politely, etc.

You may want to read "How Video Production Affects Student Engagement: An Empirical Study of MOOC Videos" (see reference [3]) as an excellent primer for deriving general wisdom from MOOC data (it explores optimal video length range for lectures). The researchers tracked 862 MOOC videos, 120,000+ students, and 6.9 million video-watching sessions on the edX platform. Now compare this MOOC-based project to an even larger system of systems. For example, the Large Hadron Collider (LHC) is the largest data production facility today. If a MOOC platform can be compared to designing a structure, then the LHC is surely an artificial palm island full of expensive houses. Each of the four experiments currently being conducted at the LHC facility produces thousands of gigabytes per second of data, which on a yearly basis results in around 15 petabytes of data. Another data monster is the Internet of Things (IoT), and you can consult reference [4] for more details.

How to Learn Data Science

This is a fundamental topic of data science, since constant learning on all levels (domain, technology, algorithm, programming language, etc.) together with practice are the hallmarks of this profession. Learning data science means gaining knowledge and understanding of data science through study, instruction, and experience. There is an ancient Chinese proverb that emphasizes the advantage of procuring knowledge over sheer consumption: "Give a man a fish and you feed him for a day. Teach a man to fish and you feed him for a lifetime." This is exactly what we strive to achieve in data science, too; instead of purely devouring current facts, we must synthesize knowledge for the future.

There are three interconnected core competency areas in data science: domain knowledge, mathematics (including probability theory and statistics), and software engineering. This triad can be conveniently illustrated with a Venn diagram, although there are a myriad of variants published since Drew Conway's original version from 2010 (you may read a funny article about these variants in reference [6]).[2] Most people are very strong in only one particular area, which is OK if they are solid in the other two. Recall that data science revolves around teamwork, which means communication capabilities at multiple levels is crucial. For example, you must speak the language of the domain to communicate properly with important stakeholders and other team members throughout the project. To make this elaboration more concrete, I will give two examples: how to acquire domain knowledge and how to acquire programming skills (see also reference [5]).

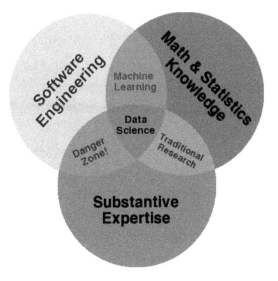

---

[2]The original version refers to *hacking skills* instead of software engineering. I think this isn't appropriate anymore. As Python data science software solutions reach an enterprise level, they must be professionally developed to be maintainable and evolvable in a cost-effective manner.

Domain Knowledge Attainment—Example

In this section, I will share my tactic to gain domain knowledge. I entered data science as a software engineer, which, as a profession, also embraces domain comprehension and mathematics. Nonetheless, data science requires a different type of acquaintance with these areas. Besides enabling effective communication, deep understanding of a domain is mandatory for the following tasks (all of them are related to data quality problems):

- Pruning of outliers (data points that are inconsistent with the rest of the dataset)

- Deleting incomplete observations

- Replacing parts of incomplete observations with estimates (for example, by a feature's mean value)

- Removing duplicates (for example, by merging records)

It is not enough to simply remove missing values. Sometimes, blind removal of data may completely distort the result. You must make a judicious decision based on domain knowledge before doing this kind of cleanup.

Suppose that you are a rookie in the field of finance and banking. You are given the task of implementing a function to calculate the growth of money. The inputs are the starting deposit (principal investment) P_0 in a bank and the annual rate of interest r (assume it is a fraction between 0 and 1). The bank uses a continuously compounded interest. You also notice a handy formula for this task, $P = P_0e^{rt}$, where t is the time in years. The final amount is P (see Exercise 1-2). Easy, right?

You should resist the temptation to immediately commence writing code without understanding all the underlying terms and mechanics. Dissecting formulas like this is the best way to peek under the hood and comprehend part of the domain (at least, from my experience). Here are the domain-related terms mentioned in the text:

- Interest rate

- Annual interest rate

- Compound interest

- Continuously compounded interest

- The mathematical constant e in the final formula

The interest rate is the percentage by which to increase the current amount. If we start with P_0, then we will end up with $P_1 = P_0 + P_0 r = P_0(1 + r)$. The annual interest rate is what we would get after one year using the previous equation. The compound interest relies on the previously accrued amount. Therefore, after 2 years we would have $P_2 = P_1 + P_1 r = P_1(1 + r) = P_0(1 + r)^2$. Notice how P_2 indirectly depends upon P_0. If a bank applies this compounding interest multiple times per year (say, with frequency n), then we would use $\dfrac{r}{n}$ as our individual interest rate and apply it subsequently n times. All in all, this would result in

$$P_1 = P_0 \left(1 + \frac{r}{n}\right)^n.$$ Finally, the continuous variant of compounding is when we let $n \to \infty$.

This is the point where we must remind ourselves from calculus that $\lim\limits_{n \to \infty}\left(1 + \dfrac{1}{n}\right)^n = \lim\limits_{t \to 0}(1+t)^{\frac{1}{t}} = e$. Consequently, we have

$$\lim_{n \to \infty}\left(1 + \frac{r}{n}\right)^n = \lim_{n \to \infty}\left(1 + \frac{1}{\frac{n}{r}}\right)^n = \lim_{n \to \infty}\left(1 + \frac{1}{\frac{n}{r}}\right)^{\frac{n}{r}r} = e^r,$$ after substituting

$t = \dfrac{1}{\frac{n}{r}} \wedge t \to 0$. If we repeat this compounding over t years, then we get our term from the initially given formula.

This technique of dissecting formulas and studying terminology is one that I always use when learning a new domain. Once you grasp all the concepts and vocabulary of a domain, then it is OK to just accept formulas. Until then, try to split up any novel (to you) descriptions from a domain into constituent parts and analyze them one by one.

Note   If you decide to use a support vector machine (SVM), then you should first read what the kernel method and kernel trick are. Similarly, before you try principal component analysis (PCA), read about eigenvectors, eigenvalues, and orthonormal basis. It is so easy to fool yourself that you've achieved something "remarkable" simply because a particular machine learning method has returned positive results.

Programming Skills Attainment—Example

Suppose that you have strong domain knowledge but not enough programming experience in Python. You may end up with a solution as shown in Listing 1-1 (for the same problem as in the previous section). Let's assume that the function should process a full list of amounts over different time periods.

Listing 1-1. Attempt to Implement a Function to Calculate the Growth of Money

```
import math

def calculate_money_growth(p0, r, t):
    # List of final amounts.
    p = []
    for i in range(len(p0)):
        p.append(p0[i] * math.exp(r * t[i]))
    return p
```

The test would consist of calling the function with p0 = [1, 2, 3], r = 0.5, and t = [1, 10, 100]. After seeing the expected output of [1.6487212707001282, 296.8263182051532, 1.5554116585761217e+22], the task would be marked as completed. Of course, for this miniature dataset, all seems to be perfect. However, when working with massive amounts of data, this approach doesn't scale. For the sake of completeness, Listing 1-2 shows the improved version utilizing NumPy. Notice the sleekness of the code and its expressive power; it speaks for itself. It also employs a basic defensive programming element.

Listing 1-2. Optimized and Safer Version of Money Growth Calculator

```
import numpy as np

def calculate_money_growth(p0, r, t):
    assert p0.size == t.size

    return p0 * np.exp(r * t)
```

Basic programming knowledge is not enough. You must be acquainted with efficient and powerful data science frameworks and technologies available for Python. Those are the true enablers to tackle Big Data problems. We will see many such frameworks in action throughout this book.

Note   At the very least, you must be proficient in SciPy, a Python-based ecosystem for data science. Moreover, as code complexity grows, you will need to know the principles, rules, and techniques for creating maintainable solutions. For example, if you've never heard of *defensive programming*, then it is time to brush up on your software engineering proficiencies. What would have happened above with p0 and t having different lengths? Where is this checked?

Overview of the Anaconda Ecosystem

Anaconda Distribution is a free ecosystem (includes a sophisticated package and environment manager) of prepackaged libraries for scientific computing and data science in Python. You may download the latest version for your operating system by visiting `https://www.anaconda.com/distribution`. There is also a paid Enterprise edition, but we will work here with the free Community variant. At the time of this writing, the current version is 5.3, which comes with Python 3.7. You may want to visit first the documentation at `https://docs.anaconda.com`.

There is also the Miniconda distribution, which contains only the bare minimum Python environment. We will use the full distribution, as we will need many bundled packages. Miniconda is good if you want to have better control over what gets installed, enabling you to save space.

After a successful installation, start Anaconda Navigator, which is a dashboard for managing environments and launching applications. Click Spyder's Launch button, as shown on Figure 1-4.

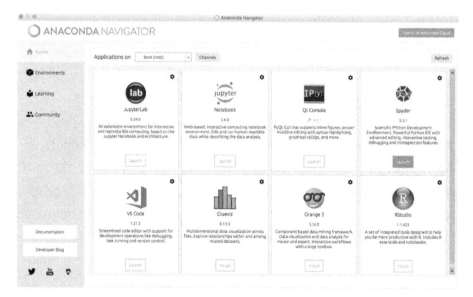

Figure 1-4. *Home page of Anaconda Navigator with the Spyder Launch button selected*

Spyder is a sophisticated scientific Python integrated development environment (IDE). It uses an internal IPython instance to execute code inside the editor and also to allow interactive computing. Figure 1-5 shows the layout of the user interface (UI).

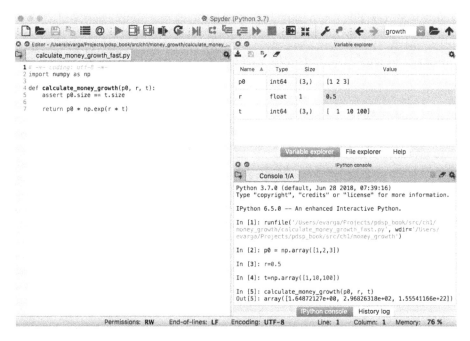

Figure 1-5. *Spyder UI with three main regions*

The Editor is located on the left side, the tabbed explorer component is in the upper-right pane (the *variable* explorer is shown in Figure 1-5), and the IPython console is in the lower-right pane, with expressions to run our fast money growth calculator. The runfile was autogenerated by Spyder after I clicked the Run button (large green arrow) on the toolbar.

If you manage to run the money growth calculator (click the Open Folder button in the toolbar to select and load a file) or evaluate any expression in the IPython console, then all is properly set up. You can also try out the embedded debugger (see the Debug menu, which is the first blue button from the left) by running some code with a configured breakpoint.

Managing Packages and Environments

conda is the command-line tool for managing packages and environments in Anaconda. You should use it instead of pip (which manages standard Python packages) and virtualenv (which creates isolated Python environments). The following is the dump of an interactive session showing the first couple of packages in the base (root) environment (observe in Figure 1-4 that Spyder was started from this environment, as shown in the Applications on field):

```
In [6]: packages = !conda list

In [7]: packages
Out[7]:
['# packages in environment at /Users/evarga/anaconda3:',
 '#',
 '# Name                    Version                   Build  Channel',
 '_ipyw_jlab_nb_ext_conf    0.1.0                     py37_0  ',
 'alabaster                 0.7.11                    py37_0  ',
 'anaconda                  5.3.0                     py37_0  ',
 'anaconda-client           1.7.2                     py37_0  ',
 'anaconda-navigator        1.9.2                     py37_0  ',
 'anaconda-project          0.8.2                     py37_0  ',
 'appdirs                   1.4.3           py37h28b3542_0  ',
 'appnope                   0.1.0                     py37_0  ',
 'appscript                 1.0.1           py37h1de35cc_1  ',
 'asn1crypto                0.24.0                    py37_0  ',
```

The first (bold) line shows an easy technique to run shell commands directly within an IPython session and capture their output. I used the list command to enumerate all packages (the central hub for Python packages is accessible at https://pypi.org). For a list of available conda commands, simply run !conda help (omit the exclamation mark if you execute conda commands from the terminal window).

The fast money growth calculator uses the numpy package, which is preinstalled in my base environment. If you do not have this package, then install it using the conda tool. Now, your solution may stick to one particular set of packages and could refuse to work on another user's machine (even if all packages are present there, too). Namely, each package has a specific version, and incompatibilities may exist between different versions. Manually ensuring that each application is run with the required set of packages is a daunting task. This is why conda has the concept of an *environment*. Think about it as a uniquely named namespace that isolates one set of packages from another.

You can manage environments from the Anaconda Navigator (select Environments in the left panel, as shown in Figure 1-4) or from the command line (using the conda tool). I will use the tool directly. To see what actions are available regarding environments, issue conda env --help and conda create --help from the shell (open a Terminal window on your operating system). First, I'll list the existing environments on my machine:

```
[~]$ conda env list
# conda environments:
#
base                     *  /Users/evarga/anaconda3
```

I will create a new environment called test with Python 3.7 and verify that it is indeed created:

```
[~]$ conda create --name test python=3.7
[~]$ conda env list
# conda environments:
#
base                     *  /Users/evarga/anaconda3
test                        /Users/evarga/anaconda3/envs/test
```

The asterisk marks the active environment (in this case, base). The test environment only contains Python 3.7 with associated packages; let's see what these are (notice that numpy is absent):

```
[~]$ conda list --name test
# packages in environment at /Users/evarga/anaconda3/envs/test:
#
# Name                    Version                   Build  Channel
ca-certificates           2018.03.07                    0
certifi                   2018.8.24                 py37_1
libcxx                    4.0.1                  h579ed51_0
libcxxabi                 4.0.1                  hebd6815_0
libedit                   3.1.20170329           hb402a30_2
libffi                    3.2.1                  h475c297_4
ncurses                   6.1                    h0a44026_0
openssl                   1.0.2p                 h1de35cc_0
pip                       10.0.1                    py37_0
python                    3.7.0                  hc167b69_0
readline                  7.0                    h1de35cc_5
setuptools                40.4.3                    py37_0
sqlite                    3.25.2                 ha441bb4_0
tk                        8.6.8                  ha441bb4_0
wheel                     0.32.0                    py37_0
xz                        5.2.4                  h1de35cc_4
zlib                      1.2.11                 hf3cbc9b_2
```

You must activate test, if you want to use it, by executing conda activate test. If you do so, then you will notice that Spyder no longer is available in Anaconda Navigator (it is offered for installation). You may now customize this environment by installing packages into it via conda install. You can customize a passive environment by specifying the --name command-line option. To deactivate an active environment, issue conda deactivate. To remove our test environment, use conda env remove --name test.

The biggest advantage of having a custom environment is that you may export it into a file and put it under version control (together with your project). Anyone may later re-create your environment by using this file. I recommend consulting the conda Cheat Sheet (see `https://docs.conda.io/projects/conda/en/latest/user-guide/cheatsheet.html`) for advice on how to perform the most frequent tasks.

Note   If you try to install a new package and receive an error, then it usually helps to update conda to the current version by running `conda update conda`. Afterward you should retry the installation.

Tip   Beware of creating a new environment by cloning an existing one. Ensure that your custom environment isn't bloated, which is something you want to avoid.

Sharing and Reproducing Environments

Data science is all about collaboration and teamwork. Sure, environments do help even in a solo arrangement; they endow you with more control over packages, thereby protecting you from inadvertently messing up a working setup. Nonetheless, their true power surfaces when others need to faithfully replicate your work. The conda tool may help here.[3]

Once you are satisfied with your environment, then you can persist it into a file by typing

```
conda env export --name SOME_ENVIRONMENT --file SOME_FILE
```

---

[3]Consult reference [8] for an overview how Docker may help package up artifacts inside containers. This a complementary approach to using Anaconda's environments and virtualization.

If you omit the environment's name, then you will save the active one. Usually, you will use `environment.yml` as a destination file (this is kind of a default name). This file should be put under version control along with any other files from your project. Anyone in a possession of this file may easily re-create your environment by issuing

```
conda create --name SOME_ENVIRONMENT --file SAVED_ENV_FILE
```

You can share Anaconda-specific files (like notebooks, environments, etc.) in many ways. Anaconda Cloud is one viable option. It allows you to share packages, environments, notebooks, and even whole projects. The latter is still in an experimental stage (at least, this was the case with the `anaconda-project` package at the time of this writing) but gives you full automation regarding project setup.

EXERCISE 1-1. TRAITS OF A DATA SCIENTIST

Suppose you must prove or disprove the claim that there are integers m and n such that $m^2 + mn + n^2$ is a perfect square (see reference [7] for a superb introduction to mathematical reasoning). How would you proceed? (Hint: Think twice before writing a single line of Python code.)

Another situation is that you are asked to write a function to sum the elements of an input floating point array with the highest degree of accuracy. You can neglect the performance implications. What would be the winning approach? If you have managed to solve the problem, then try to explain why the function is behaving as such. Have you ever wondered what causes rounding errors? How is the `float` data type represented in Python?

A data scientist must possess good analytical skills, be creative, use common sense, and cultivate an inclination to ask *Why?* questions.

EXERCISE 1-2. NAMING ABSTRACTIONS

The description and the associated code that follows use strange names P_0 and P. In finance there are established terms for these values: present value (PV) and future value (FV), respectively. Alter the associated money growth calculator to use these established terms instead. Do you agree that this simple action has improved comprehensibility?

One of the best ways to check whether your design is sound is by looking at the names of your abstractions. All of them must have a meaningful name. Otherwise, you should rethink the structure of your code. Don't treat naming as something nobody cares about (your code will be read and maintained by humans).

Summary

Data science is a highly collaborative field that follows an iterative and incremental process. This process consists of many phases, such as planning, data engineering (acquisition, preprocessing, exploration), computational data analysis, reporting, and execution of actions. The aim is to either generate actionable insights from data or explain past phenomena (a.k.a. root cause analysis). A data scientist may play different roles throughout the project, and should possess adequate domain knowledge, math, and software engineering skills. Python is a beloved language of data science because it offers powerful frameworks (besides being a modern and flexible programming language supporting both object-oriented and functional styles).

Anaconda is the recommended ecosystem to develop data science solutions. It bundles all the necessary applications (such as JupyterLab, Spyder, Glueviz, Orange 3, etc.), tools (like the conda manager), and data science packages into a cohesive system. Anaconda also promotes the polyglot programming paradigm, allowing you to implement parts of your project's portfolio in a multitude of programming languages. It also solves the common problem of crafting reproducible solutions by supporting sophisticated management of packages and environments.

References

1. Suzanne Robertson and James Robertson, "How Now Brown Cow," *The Atlantic Systems Guild*, `https://www.volere.org/wp-content/uploads/2019/02/howNowBrownCow.pdf`, May 2009.

2. Ani Adhikari and John DeNero, *Computational and Inferential Thinking: The Foundations of Data Science*, `https://www.inferentialthinking.com`, 2015.

3. Philip J. Guo, Juho Kim, and Rob Rubin, "How Video Production Affects Student Engagement: An Empirical Study of MOOC Videos," *Proceedings of the First ACM Conference on Learning @ Scale Conference (L@S '14)*, pp. 41–50 (available at `http://pgbovine.net/publications/edX-MOOC-video-production-and-engagement_LAS-2014.pdf`).

4. Ervin Varga, Draško Drašković, and Dejan Mijić, *Scalable Architecture for the Internet of Things: An Introduction to Data-Driven Computing Platforms*, O'Reilly, 2018.

5. Sinan Ozdemir, *Principles of Data Science*, Packt Publishing, 2016.

6. David Taylor, "Battle of the Data Science Venn Diagrams," https://www.kdnuggets.com/2016/10/battle-data-science-venn-diagrams.html, Oct. 6, 2016.

7. Keith Devlin, *Introduction to Mathematical Thinking*, Keith Devlin, 2012.

8. Joshua Cook, *Docker for Data Science: Building Scalable and Extensible Data Infrastructure Around the Jupyter Notebook Server*, Apress, 2017.

CHAPTER 2

Data Engineering

After project initiation, the data engineering team takes over to build necessary infrastructure to acquire (identify, retrieve, and query), munge, explore, and persist data. The goal is to enable further data analysis tasks. Data engineering requires different expertise than is required in later stages of a data science process. It is typically an engineering discipline oriented toward craftsmanship to provide necessary input to later phases. Often disparate technologies must be orchestrated to handle data communication protocols and formats, perform exploratory visualizations, and preprocess (clean, integrate, and package), scale, and transform data. All these tasks must be done in context of a global project vision and mission relying on domain knowledge. It is extremely rare that raw data from sources is immediately in perfect shape to perform analysis. Even in the case of a clean dataset, there is often a need to simplify it. Consequently, dimensionality reduction coupled with feature selection (remove, add, and combine) is also part of data engineering. This chapter illustrates data engineering through a detailed case study, which highlights most aspects of it. The chapter also presents some publicly available data sources.

Data engineers also must be inquisitive about data collection methods. Often this fact is neglected, and the focus is put on data representation. It is always possible to alter the raw data format, something hardly imaginable regarding data collection (unless you are willing to redo the whole data acquisition effort). For example, if you receive a survey result in an Excel file, then it is a no-brainer to transform it and save it in a relational

© Ervin Varga 2019
E. Varga, *Practical Data Science with Python 3*,
https://doi.org/10.1007/978-1-4842-4859-1_2

database. On the other hand, if the survey participants were not carefully selected, then the input could be biased (favor a particular user category). You cannot apply a tool or program to correct such a mistake.

Figure 2-1 shows the well-known golden principle of data preparation: if you get garbage on input, you're going to have garbage output.

Figure 2-1. *Poorly worded questions do not lead to good answers. Similarly, bad input cannot result in good output.*

E-Commerce Customer Segmentation: Case Study

This case study introduces exploratory data analysis (EDA) by using the freely available small dataset from `https://github.com/oreillymedia/doing_data_science` (see also [1] in the "References" section at the end of the chapter).[1] It contains (among other data) simulated observations about ads shown and clicks recorded on the New York Times home page in May 2012. There are 31 files in CSV format, one file per day; each file is named `nyt<DD>.csv`, where `DD` is the day of the month (for example, `nyt1.csv` is the file for May 1). Every line in a file designates a single user. The following characteristics are tracked: age, gender (0=female, 1=male), number of

---

[1]Very large datasets shouldn't be kept in a Git repository. It is better to store them in a cloud (S3, Google Drive, Dropbox, etc.) and download from there.

impressions, number of clicks, and signed-in status (0=not signed in, 1=signed in).

The preceding description is representative of the usual starting point in many data science projects. Based on this description, we can deduce that this case study involves *structured data* (see also the sidebar "Flavors of Data") because the data is organized in two dimensions (rows and columns) following a CSV format, where each column item is single valued and mandatory. If the description indicated some column items were allowed to take multiple values (for example, a list of values as text), be absent, or change data type, then we would be dealing with *semistructured data*. Nevertheless, we don't know anything about the quality of the content nor have any hidden assumptions regarding the content. This is why we must perform EDA: to acquaint ourselves with data, get invaluable perceptions regarding users' behavior, and prepare everything for further analysis. During EDA we can also notice data acquisition deficiencies (like missing or invalid data) and report these findings as defects in parallel with cleaning activities.

Note   *Structured data* doesn't imply that the data is tidy and accurate. It just emphasizes what our expectation is about the organization of the data, estimated data processing effort, and complexity.

FLAVORS OF DATA

We have two general categories of data types: *quantitative* (numeric) and *qualitative* (categorical). You can apply meaningful arithmetic operations only to quantitative data. In our case study, the number of impressions is a quantitative variable, while gender is qualitative despite being encoded numerically (we instead could have used the symbols F and M, for example).

A data type may be on one of the following levels: nominal, ordinal, interval, or ratio. *Nominal* is a bare enumeration of categories (such as gender and logged-in status). It makes no sense to compare the categories nor perform mathematical operations. The *ordinal* level establishes ordering of values (T-shirt sizes XS, S, M, L, and XL are categorical with clear order regarding size). *Interval* applies to quantitative variables where addition and subtraction makes sense. Nonetheless, multiplication and division doesn't (this means there is no proper starting point for values). The *ratio* level allows all arithmetic operations, although in practice these variables are restricted to be non-negative. In our case study, age, number of impressions, and number of clicks are all at this level (all of them are non-negative).

Understanding the differentiation of levels is crucial to understanding the appropriate statistics to summarize data. There are two basic descriptors of data: *center of mass* describes where the data is *tending to,* and *standard deviation* describes the average spread around a center (this isn't a precise formulation, but will do for now). For example, it makes no sense to find the mean of nominal values. It only makes sense to dump the mode (most frequent item). For ordinals, it makes sense to present the median, while for interval and ratio we usually use the mean (of course, for them, all centers apply).

Chapter 1 covered the data science process and highlighted the importance of project initiation, the phase during which we establish a shared vision and mission about the scope of the work and major goals. The data engineering team must be aware of these and be totally aligned with the rest of the stakeholders on the project. This entails enough domain knowledge to open up proper communication channels and make the outcome of the data engineering effort useful to the others. Randomly poking around is only good to fool yourself that you are performing some "exploration." All actions must be framed by a common context.

That being said, let us revisit the major aims of the case study here and penetrate deeper into the marketing domain. The initial data collection effort was oriented toward higher visibility pertaining to advertisement efforts. It is quite common that companies just acquire data at the beginning, as part of their strategic ambitions. With this data available, we now need to take the next step, which is to improve sales by better managing user campaigns.

As users visit a web site, they are exposed to various ads; each exposure increases the *number of impressions* counter. The goal is to attract users to click ads, thus making opportunities for a sale. A useful metric to trace users' behavior is the *click-through rate* (part of domain knowledge), which is defined as $\dfrac{number\ of\ clicks}{number\ of\ impressions}$. On average, this rate is quite small (2% is regarded as a high value), so even a tiny improvement means a lot. The quest is to figure out what is the best tactic to increase the click-through rate.

It is a well-known fact in marketing that customized campaigns are much more effective than just blindly pouring ads on users. Any customization inherently assumes the existence of different user groups; otherwise, there would be no basis for customization. At this moment, we cannot make any assumptions about campaign design nor how to craft groups. Nonetheless, we can try to research what type of segmentation shows the greatest variation in users' behavior. Furthermore, we must identify what are the best indicators to describe users. All in all, we have a lot of moving targets to investigate; this shouldn't surprise you, as we are in the midst of EDA. Once we know where to go, then we could automate the whole process by creating a full data preparation pipeline (we will talk about such pipelines in later chapters).

Creating a Project in Spyder

Launch Spyder (refer to Chapter 1 for the Anaconda setup instructions, if needed) and select the Projects ➤ New Project menu item. In the dialog box that opens, specify the project's name (in our case, segmentation), location (choose a folder on your machine to become this project's parent directory), and type (leave it as `Empty Project`). In the File Explorer, create a folder structure as shown in Figure 2-2. The `raw_data` folder contains the unpacked files, the `results` folder hosts the cleaned and prepared data, and the `scripts` folder stores the various Python programs. The `segmentation` folder contains the driver code. In subsequent sections you will need to create Python files by selecting File ➤ New file.

Figure 2-2. *Folder structure of our Spyder project*

Downloading the Dataset

You should now download the dataset and unpack it. You may do this manually, via shell script, or from Python (to name a few options). I've chosen the last option to demonstrate a couple of techniques in Python, although I do prefer bash scripts for this stuff. Listing 2-1 shows the download script, `nyt_data.py`, which downloads the dataset and performs necessary housekeeping. Notice how it unpacks the embedded ZIP file containing our data. The `cleanup` procedure showcases basic exception handling.

Listing 2-1. The nyt_data.py Script

```
import requests, zipfile, io, shutil

unpackedFolder = '/dds_datasets/'
unpackedZipFile = 'dds_ch2_nyt.zip'
```

```
def retrieve(sourceFile, destinationFolder):
    def cleanup():
        try:
            shutil.rmtree(destinationFolder + unpackedFolder)
        except OSError as e:
            print("Folder: %s, Error: %s" % (e.filename,
            e.strerror))

    r = requests.get(sourceFile)
    assert r.status_code == requests.codes.ok

    z = zipfile.ZipFile(io.BytesIO(r.content))
    z.extractall(destinationFolder)

    # The top archive contains another ZIP file with our data.
    z = zipfile.ZipFile(destinationFolder + unpackedFolder +
    unpackedZipFile)
    z.extractall(destinationFolder)

    cleanup()
```

The requests package is the main HTTP library for Python (see https://requests.readthedocs.io). The retrieved content is available under r.content. The zipfile package manages archives, and shutil provides high-level file operations. We use it to remove the temporary folder with embedded archives. Listing 2-2 shows the driver code for this part of the job. It is the first part of the driver.py script for downloading the raw data. Since it calls scripts in the scripts folder, this must be added to the path. The driver must be executed from the project's base folder. Also, observe that the fileUrl starts with raw instead of blob.

Listing 2-2. The Driver Code

```
import sys
import os
sys.path.append(os.path.abspath('scripts'))

from nyt_data import retrieve

repoUrl = 'https://github.com/oreillymedia/doing_data_science/'
fileUrl = 'raw/master/dds_datasets.zip'

retrieve(repoUrl + fileUrl, 'raw_data')
print('Raw data files are successfully retrieved.')
```

After the execution of Listing 2-2, the CSV files will be available in the raw_data subfolder. You can easily verify this in Spyder's File Explorer.

Exploring the Dataset

Let's start our journey by peeking into our dataset. We will carry out all steps from Spyder's IPython console. This provides an interactive working environment, and this is the cornerstone of why Python is so popular among data scientists. You can issue statements and watch their immediate execution. There's no need to wait for long edit/build/execute cycles as is the case with compiled languages. Below is the result of printing the current working directory (your output will differ depending on your project's parent folder)[2]:

```
>> %pwd
'/Users/evarga/Projects/pdsp_book/src/ch2/segmentation'
```

---

[2]I have omitted the In[...] and Out[...] prompts for brevity and just marked the input prompt by >>. Also, keep in mind that Tab completion works for all parts of a command, including file names. Just press Tab and see what Spyder offers to you.

The %pwd magic command is mimicking the native pwd bash
command. You could also get the same result with !pwd (! is the escape
character to denote what follows is a shell command). The shell-related
magic commands are operating system neutral, so it makes sense to
use them when available. Sometimes, as is the case with !cd, the native
shell command cannot even be properly run (you must rely on the
corresponding magic variant). Visit https://ipython.readthedocs.io/
en/stable/interactive/magics.html to receive help on built-in magic
commands; run %man <shell command> to get help on a shell magic
command (for example, try %man head or %man wc). Now, switch into the
raw_data folder by running %cd raw_data (watch how the File Explorer
refreshes its display).

I advise you to always at first preview some raw data file using the most
rudimentary technique. Once you develop your first impression of what
is there, then you can utilize more powerful tools and frameworks for data
manipulation. Here is the dump of the nyt1.csv file (only the first ten lines):

```
>> !head -n 10 nyt1.csv
"Age","Gender","Impressions","Clicks","Signed_In"
36,0,3,0,1
73,1,3,0,1
30,0,3,0,1
49,1,3,0,1
47,1,11,0,1
47,0,11,1,1
0,0,7,1,0
46,0,5,0,1
16,0,3,0,1
```

The first line is a header with decently named columns. The rest of the
lines are user records. Nonetheless, just with these ten lines displayed, we
can immediately spot a strange value for age: 0. We don't expect it be non-
positive. We will need to investigate this later.

By looking at file sizes in the File Explorer, we see that they are all around 4.5MB (a couple of them are larger, between 7.5MB and 8MB). Consequently, most of them should contain a similar number of records. Let's find out how many records are in the smallest file and in the biggest file (click the Size tab to sort files by their size):

```
>> !wc -l nyt26.csv
362235 nyt26.csv
```

```
>> !wc -l nyt13.csv
786045 nyt13.csv
```

So, the number of records fluctuates in the range of [360K, 790K] with sizes between 3.7MB and 8.0MB; we shouldn't have a problem processing a single file in memory even on a modest laptop.

Finding Associations Between Features

The previously mentioned columns are our basic features for describing each web page visitor. Before we delve deeper into any feature engineering (a process of creating more informative characteristics, like the already mentioned click-through rate), we must see how the existing features interplay. Any rule or pattern designates a particular association (relationship). Visualization is very effective here and complements descriptive statistics (like minimum, maximum, mean, etc.). The SciPy ecosystem offers the Pandas framework to work with tabular data, including visualization, which is just what we need here (see reference [3] for superb coverage of Pandas). Recall that we are in the EDA stage, so nothing fancy is required (Chapter 6 describes more advanced visualizations). Type the following snippet in your console:

```
>> import pandas as pd
```

```
>> nyt_data = pd.read_csv('nyt1.csv')
>> nyt_data.head(n=10)
```

	Age	Gender	Impressions	Clicks	Signed_In
0	36	0	3	0	1
1	73	1	3	0	1
2	30	0	3	0	1
3	49	1	3	0	1
4	47	1	11	0	1
5	47	0	11	1	1
6	0	0	7	1	0
7	46	0	5	0	1
8	16	0	3	0	1
9	52	0	4	0	1

The pd is a common acronym for the pandas package (like np is for numpy). The read_csv function does all the gory details of parsing the CSV file and forming the appropriate DataFrame object. This object is the principal entity in the Pandas framework for handling 2D data. The head method shows the first specified number of records in a nice tabular format. Observe that our table has inherited the column names from the CSV file. The leftmost numbers are indices; by default, they are record numbers starting at 0, but the example below shows how it may be easily redefined.

The next line shows the overall dimensionality of our table:

```
>> nyt_data.shape
(458441, 5)
```

We have 458441 records (rows) and 5 features (columns). The number of rows is very useful for finding out missing column values. So, you should always print the shape of your table.

The next snippet presents the overall descriptive statistics for columns (watch out for the custom index):

```
>> summary_stat = nyt_data.describe()
>> summary_stat
                Age            Gender   ...            Clicks        Signed_In
count   458441.000000   458441.000000  ...     458441.000000    458441.000000
mean        29.482551        0.367037  ...          0.092594         0.700930
std         23.607034        0.481997  ...          0.309973         0.457851
min          0.000000        0.000000  ...          0.000000         0.000000
25%          0.000000        0.000000  ...          0.000000         0.000000
50%         31.000000        0.000000  ...          0.000000         1.000000
75%         48.000000        1.000000  ...          0.000000         1.000000
max        108.000000        1.000000  ...          4.000000         1.000000

[8 rows x 5 columns]
```

```
>> summary_stat['Impressions']
count    458441.000000
mean          5.007316
std           2.239349
min           0.000000
25%           3.000000
50%           5.000000
75%           6.000000
max          20.000000
Name: Impressions, dtype: float64
```

The describe method calculates the descriptive statistics (as shown in the leftmost part of the output) for each column and returns the result as a new DataFrame. When we display summary_stat, then it automatically abbreviates the output if all columns cannot fit on the screen. This is why we need to explicitly print the information for the Impressions column.

At first glance, all counts equal the total number of rows, so we have no missing values (at least, in this file). The Age column's minimum value is zero, although we have already noticed this peculiarity. About 36% of users are male and 70% of them were signed in (the mean is calculated by summing up the 1s and dividing the sum by the number of rows). Well, presenting a mean for a categorical variable isn't right, but Pandas has no way to know what terms like Gender or Signed In signify. Apparently, conveying semantics is much more than just properly naming things.

The next excerpt checks what Pandas thinks about column types:

```
>> nyt_data.dtypes
Age             int64
Gender          int64
Impressions     int64
Clicks          int64
Signed_In       int64
dtype: object
```

Gender and Signed_In are not ratio-level values, although in many scenarios using numerical encoding for Boolean values is beneficial (they don't confuse many machine-learning algorithms). The procedure to make quantitative variables out of qualitative ones is called *making dummy variables* (see also the pandas.get_dummies function). Let's help Pandas and mark our nominal columns properly:

```
>> nyt_data['Gender'] = nyt_data['Gender'].astype('category')
>> nyt_data['Signed_In'] = nyt_data['Signed_In'].
astype('category')
>> nyt_data['Gender'].describe()
count      458441
unique          2
top             0
freq       290176
Name: Gender, dtype: int64
```

The first (bold) line shows how to change the column's type to category. Look how the type of statistics was altered to make sense for a categorical variable. We will later bundle up all these findings in a script to automate everything. Currently, we are solely playing with data. Think again how interactivity helps us in this mission. The next two lines produce two scatter plots, Age vs. Impressions and Age vs. Clicks, as shown in Figure 2-3 (a semicolon at the end of a line prevents printing the function's return value):

```
>> nyt_data.plot(x='Age', y='Impressions', c=nyt_
data['Gender'], kind='scatter', ➡
colormap='Paired', colorbar=False);
>> nyt_data.plot(x='Age', y='Clicks', c=nyt_data['Gender'],
kind='scatter', ➡
colormap='Paired', colorbar=False);
```

Since plotting in Python is so easy, you should always make some plots. There is no guarantee that you will immediately hit the mark, but you could learn something. For example, we have discovered that age zero isn't a bare error in data (all users are designated as females, cover the full range of impressions and clicks, and there is a distinct gap from the non-zero group). The next snippet selects users with age zero:

```
>> zero_age_users = nyt_data[nyt_data['Age'] == 0]
>> zero_age_users.head(n=10)
```

	Age	Gender	Impressions	Clicks	Signed_In
6	0	0	7	1	0
10	0	0	8	1	0
12	0	0	4	0	0
15	0	0	6	0	0
19	0	0	5	0	0
24	0	0	4	0	0

39	0	0	7	2	0
41	0	0	4	0	0
46	0	0	3	0	0
47	0	0	7	0	0

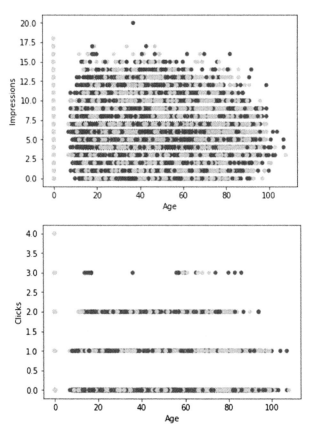

Figure 2-3. *A scatter plot is handy to show potential relationships between numerical variables*

You can add a third dimension by differently coloring the points.[3] The dark dots denote males, while the light ones denote females. Due to the very large number of records (points are heavily overlapped), we cannot discern any clear pattern except that the zero age group is special.

The bolded section illustrates the *Boolean indexing* technique (a.k.a. *masking*), where a Boolean vector is used for filtering records. Only those records are selected where the matching element is True. Based on the dump, we see that all users are marked as females and logged out.

The next lines assure that zero age actually designates a logged-out category:

```
>> logged_in_zero_age_users_mask = (nyt_data['Age'] == 0) &
(nyt_data['Signed_In'] == 1)
>> logged_in_zero_age_users_mask.any()
False

>> logged_out_non_zero_age_users_mask = (nyt_data['Age'] > 0) &
(nyt_data['Signed_In'] == 0)
>> logged_out_non_zero_age_users_mask.any()
False
```

The any method checks if there is at least one True element in a Boolean mask. Both times we see False. Therefore, we can announce a very important rule that every logged-out user is signaled with zero age, which absolutely makes sense. If a user isn't logged in, then how could we know her age? Observe that we have gradually reached this conclusion, always using data as a source of truth buttressed with logic and domain knowledge. All in all, the Signed_In column seems redundant (noninformative), but it may be useful for grouping purposes.

---

[3]Visit https://matplotlib.org/examples/color/colormaps_reference.html to browse the available colormaps. Each sample is named (find the one that we have used here).

Don't worry too much at this point about performance or elegancy. Many solutions are possible for the same problem. For example, here is another way to check that all logged-out users are of age zero:

```
>> grouped_by_logged_status = nyt_data.groupby('Signed_In')
>> import numpy as np
>> grouped_by_logged_status.agg([np.min, np.max])['Age']
          amin   amax
Signed_In
0             0      0
1             7    108
```

We have grouped the data by the signed-in status and aggregated it based on minimum and maximum. For a logged-out state, the only possible age value is zero. For a logged-in state, the youngest user is 7 years old.

Incorporating Custom Features

Feature engineering is both art and science that heavily relies on domain knowledge. Many scientists from academia and industry constantly seek new ways to characterize observations. Most findings arc published, so that others may continue to work on various improvements. In the domain of e-commerce, it turned out that click-through rate, demographics (in our case study, gender), and age groups (see reference [2]) are informative features.

Let's start by introducing a new column, Age_Group, that classifies users as "Unknown", "1-17", "18-24", "25-34", "35-44", "45-54", "55-64", and "65+" (we remove the Age column):

```
>> nyt_data['Age_Group'] = pd.cut(nyt_data['Age'], bins=[-1, 0,
17, 24, 34, 44, 54, 64, 120], ➧
labels=["Unknown", "1-17", "18-24", "25-34", "35-44", "45-54",
"55-64", "65+"])
>> nyt_data.drop('Age', axis='columns', inplace=True)
```

```
>> nyt_data.head(n=10)
```

	Gender	Impressions	Clicks	Signed_In	Age_Group
0	0	3	0	1	35-44
1	1	3	0	1	65+
2	0	3	0	1	25-34
3	1	3	0	1	45-54
4	1	11	0	1	45-54
5	0	11	1	1	45-54
6	0	7	1	0	Unknown
7	0	5	0	1	45-54
8	0	3	0	1	1-17
9	0	4	0	1	45-54

```
>> nyt_data['Age_Group'].dtypes
CategoricalDtype(categories=['Unknown', '1-17', '18-24', '25-34',
'35-44', '45-54',
                '55-64', '65+'],
            ordered=True)
```

The following code adds the click-through rate (CTR) column to our table:

```
>> nyt_data['CTR'] = nyt_data['Clicks'] / nyt_data['Impressions']
```

```
>> nyt_data.head(n=10)
```

	Gender	Impressions	Clicks	Signed_In	Age_Group	CTR
0	0	3	0	1	35-44	0.000000
1	1	3	0	1	65+	0.000000
2	0	3	0	1	25-34	0.000000
3	1	3	0	1	45-54	0.000000
4	1	11	0	1	45-54	0.000000
5	0	11	1	1	45-54	0.090909

6	0	7	1	0	Unknown	0.142857
7	0	5	0	1	45-54	0.000000
8	0	3	0	1	1-17	0.000000
9	0	4	0	1	45-54	0.000000

We have to be careful to avoid potential division by zero issues. Pandas will automatically mark such values as *Not a Number (NaN)*. We could instantly remove such rows, but we must be careful here. The lack of ads is a sign that potential buyers are omitted from a campaign. The next statement shows how many users haven't seen ads (True values are treated as 1s and False values as 0s, so we can directly sum them up):

```
>> np.sum(nyt_data['Impressions'] == 0)
3066
```

This is a low percentage of users (about 0.6%). Removing them wouldn't impact the analysis. The next statement prunes away rows whose CTR is NaN:

```
>> nyt_data.dropna(inplace=True)
```

The click-through rate is a number from the interval [0, 1]. Consequently, we should delete all rows where the number of clicks is higher than the number of impressions. The next step is kind of a generic sanity action that we will apply to all data files:

```
>> nyt_data.drop((nyt_data['Clicks'] >
nyt_data['Impressions']).nonzero()[0], inplace=True)
```

The Boolean mask must be converted to an array of indices where elements are True. This is the job of the nonzero method. Of course, we could also get an empty array, but this is OK. As you likely realize, besides exploring, we are also cleaning up the dataset. Most actions in data science are intermixed, and it is hard to draw borders between activity spaces.

The following lines produce two bar plots to visualize the distributions of the number of impressions and CTR for different age ranges (these are shown in Figure 2-4):

```
>> grouped_by_age_group = nyt_data.groupby(by='Age_Group').
agg(np.mean)
>> grouped_by_age_group['Impressions'].plot(kind='bar',
colormap='tab20c', title='Impressions');
>> grouped_by_age_group['CTR'].plot(kind='bar',
colormap='tab20c', title='CTR');
```

Another interesting indicator may be the total number of clicks across age groups and logged-in statuses. The following lines produce two bar plots to visualize these metrics (these are shown in Figure 2-5):

```
>> grouped_by_age_group = nyt_data.groupby(by='Age_Group').
agg(np.sum)
>> grouped_by_logged_status = nyt_data.groupby(by='Signed_In').
agg(np.sum)
>> grouped_by_age_group['Clicks'].plot(kind='bar',
colormap='tab20c', title='Clicks');
>> grouped_by_logged_status['Clicks'].plot(kind='bar',
colormap='tab20c', title='Clicks');
```

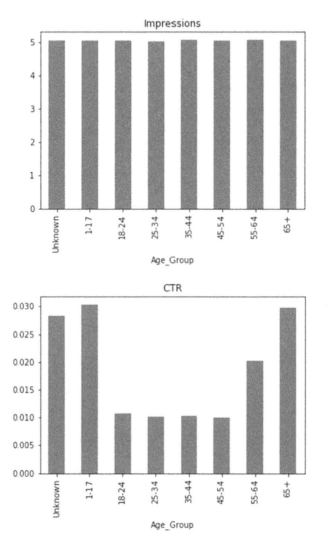

Figure 2-4.   *These two bar plots clearly emphasize the difference between features: number of impressions and CTR. The youngest and oldest people are more susceptible to click ads. The unknown age group is a mixture of all ages. At any rate, the segmentation by age groups turns out to be useful.*

It is an amazing fact that 30% of unregistered users generate the same total number of clicks as the 70% registered ones. Furthermore, we can notice a steady increase in the number of clicks as age ranges advance. This might suggest that older people have more money to spend on shopping than younger ones. It could also be the case that older people are more attracted to the site than youngsters. Investigating the reason at some point in the future makes sense. At this moment, we are simply trying out various approaches to best describe users' behavior.

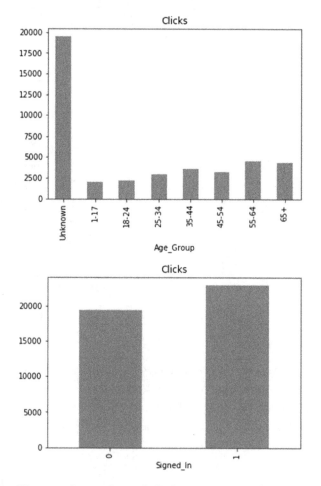

Figure 2-5. *The total number of clicks represents the matching group's importance (potential) to generate sale*

Figure 2-6 illustrates how click-through rate and total number of clicks are distributed across combinations of age ranges and demographics. The next couple of lines implement this idea:

```
>> grouped_by_age_group_and_gender = nyt_data.groupby(by=
['Age_Group', 'Gender'])➡
[['CTR', 'Clicks']].agg([np.mean, np.sum])
>> grouped_by_age_group_and_gender['CTR']['mean'].
plot(kind='bar', colormap='tab20c',➡ title='CTR', rot=45);
>> grouped_by_age_group_and_gender['Clicks']['sum'].
plot(kind='bar', colormap='tab20c',➡ title='Clicks', rot=45);
```

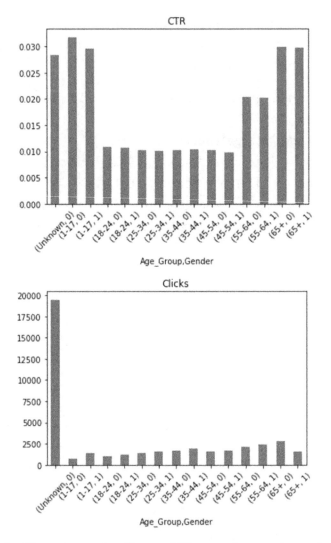

Figure 2-6. *There is no significant difference between genders, aside from total number of clicks for the youngest and oldest age groups*

These diagrams also show how easy it is to customize plots. To enhance readability, I have printed the X axis labels slanted by 45 degrees.

We are now ready to capture our findings in tabular format; it compresses the whole data file into a set of descriptive statistics while still preserving crucial traits of user behavior. This is the essence of data preprocessing by means of mathematical statistics. The next snippet produces a multilevel indexed data frame:

```
>> def q25(x):
>>      return x.quantile(0.25)

>> def q75(x):
>>      return x.quantile(0.75)

>> compressed_nyt_data = nyt_data.groupby(by=['Age_Group',
'Gender'])➡
[['CTR', 'Clicks']].agg([np.mean, np.std, np.max, q25,
np.median, q75, np.sum])
>> compressed_nyt_data
```

		CTR				...	Clicks			
		mean	std	amax	q25	...	q25	median	q75	sum
Age_Group	Gender					...				
Unknown	0	0.028355	0.085324	1.0	0.0	...	0	0	0	19480
1-17	0	0.031786	0.088829	1.0	0.0	...	0	0	0	683
	1	0.029582	0.087169	1.0	0.0	...	0	0	0	1382
18-24	0	0.010789	0.053556	1.0	0.0	...	0	0	0	1002
	1	0.010685	0.052763	1.0	0.0	...	0	0	0	1165
25-34	0	0.010186	0.049832	1.0	0.0	...	0	0	0	1388
	1	0.010111	0.051750	1.0	0.0	...	0	0	0	1549
35-44	0	0.010234	0.050691	1.0	0.0	...	0	0	0	1707
	1	0.010332	0.051503	1.0	0.0	...	0	0	0	1955
45-54	0	0.010187	0.050994	1.0	0.0	...	0	0	0	1542
	1	0.009754	0.049075	1.0	0.0	...	0	0	0	1690
55-64	0	0.020378	0.072637	1.0	0.0	...	0	0	0	2105
	1	0.020245	0.071184	1.0	0.0	...	0	0	0	2451
65+	0	0.029856	0.084455	1.0	0.0	...	0	0	0	2765
	1	0.029709	0.083338	1.0	0.0	...	0	0	0	1585

```
[15 rows x 14 columns]
>> compressed_nyt_data['Clicks']
```

Age_Group	Gender	mean	std	amax	q25	median	q75	sum
Unknown	0	0.143049	0.386644	4	0	0	0	19480
1-17	0	0.157664	0.386004	2	0	0	0	683
	1	0.147021	0.383844	3	0	0	0	1382
18-24	0	0.053207	0.230291	2	0	0	0	1002
	1	0.053995	0.233081	2	0	0	0	1165
25-34	0	0.051310	0.225113	2	0	0	0	1388
	1	0.050374	0.224588	2	0	0	0	1549
35-44	0	0.051782	0.225929	2	0	0	0	1707
	1	0.052232	0.227133	3	0	0	0	1955
45-54	0	0.051303	0.227742	2	0	0	0	1542
	1	0.050018	0.224011	2	0	0	0	1690
55-64	0	0.102026	0.318453	3	0	0	0	2105
	1	0.102854	0.320578	3	0	0	0	2451
65+	0	0.152065	0.384822	3	0	0	0	2765
	1	0.152801	0.386677	3	0	0	0	1585

The preceding table contains enough information to even restore some other properties. For example, the size of groups may be calculated by dividing the corresponding total number of clicks by its mean.

Automating the Steps

Listing 2-3 shows the extension of the script nyt_data.py with a summarize function that gathers all the pertinent steps from our exploration. It receives a data file name and returns a DataFrame instance containing descriptive statistics for depicting user behavior.

Listing 2-3. The summarize Function Responsible for Preprocessing
a Single Data File

```
import pandas as pd
import numpy as np

def summarize(data_file):
    def q25(x):
        return x.quantile(0.25)

    def q75(x):
        return x.quantile(0.75)

    # Read and parse the CSV data file.
    nyt_data = pd.read_csv(data_file, dtype={'Gender':
    'category'})

    # Segment users into age groups.
    nyt_data['Age_Group'] = pd.cut(
            nyt_data['Age'],
            bins=[-1, 0, 17, 24, 34, 44, 54, 64, 120],
            labels=["Unknown",
                    "1-17",
                    "18-24",
                    "25-34",
                    "35-44",
                    "45-54",
                    "55-64",
                    "65+"])
    nyt_data.drop('Age', axis='columns', inplace=True)
    # Create the click through rate feature.
    nyt_data['CTR'] = nyt_data['Clicks'] / nyt_
    data['Impressions']
    nyt_data.dropna(inplace=True)
```

```
nyt_data.drop((nyt_data['Clicks'] > nyt_data
['Impressions']).nonzero()[0],
                inplace=True)

# Make final description of data.
compressed_nyt_data = \
    nyt_data.groupby(by=['Age_Group', 'Gender'])[['CTR',
    'Clicks']] \
            .agg([np.mean, np.std, np.max, q25, np.median,
            q75, np.sum])
return compressed_nyt_data
```

You should get back the same result as from the previous section after passing the 'nyt1.csv' file to summarize. We are now left with implementing the pipeline to process all files. Observe how all those preliminary sidesteps and visualizations are gone. The source code is purely an end result of complex intellectual processes.

Inspecting Results

The next stage is to create a test pipeline to process all files. Parts of this test harness will transition into the final data pipeline infrastructure. Listing 2-4 shows the last extension to the nyt_data.py script. The traverse function calls back a custom collector, which further handles the data frame instance. The function uses the pathlib package introduced in Python 3.4. The get_file_number function extracts the sequence number from a data file name. This is important, as files may be traversed in arbitrary order. The file number is used to index into a collection storing some values from each data frame.

Listing 2-4. The `traverse` Function Calling Back a Custom Collector

```
import pathlib

def traverse(sourceFolder, collect):
    def get_file_number(data_file):
        return int(data_file.name[3:-4]) - 1

    for data_file in pathlib.Path(sourceFolder).
    glob('nyt*.csv'):
        collect(summarize(data_file.absolute()), get_file_
        number(data_file))
```

Listing 2-5 shows the section from `driver.py` to call `traverse` and make some plots as shown in Figure 2-7 (those diagrams are vital to reinforce the expressive power of our segmentation tactic and accumulated statistics). Notice how `traverse` is called with the `select_stats_unregistered` callback function to collect pertinent data for each file in Listing 2-5. The main program collects the average click-through rate and total clicks for unregistered users that are representative enough for postulating a hypothesis.

Listing 2-5. `np.empty` Allocates Spaces for 31 Days of Data Points

```
import numpy as np
import pandas as pd
from nyt_data import traverse

summary_data = dict()
summary_data.setdefault('CTR', np.empty(31))
summary_data.setdefault('Clicks', np.empty(31))

def select_stats_unregistered(df, file_num):
    summary_data['CTR'][file_num] = df['CTR']['mean']
    [('Unknown', '0')]
```

```python
    summary_data['Clicks'][file_num] = df['Clicks']['sum']
    [('Unknown', '0')]

traverse('raw_data', select_stats_unregistered)
print('Raw data files are successfully processed.')

# Make some plots of CTR and Total Clicks over time.
df = pd.DataFrame.from_dict(summary_data)

import matplotlib.pyplot as plt

fig, axes = plt.subplots(nrows=2, ncols=1)
df['CTR'].plot(
        title='Click Through Rate Over 1 Month',
        ax=axes[0],
        figsize=(8, 9),
        xticks=[]
);
df['Clicks'].plot(
        xticks=range(0, 31, 2),
        title='Total Clicks Over 1 Month',
        ax=axes[1],
        figsize=(8, 9)
);
```

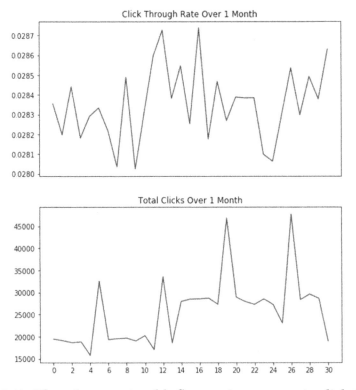

Figure 2-7. *There is no noticeable fluctuation pattern in click-through rate*

Notice that there are clear peaks in the total number of clicks. They happened on the 6th, 13th, 20th, and 27th of May 2012, which are all Sundays (the numbering starts at zero; see also Exercise 2-2). Consequently, on those days the web site had experienced much higher traffic than during weekdays. Moreover, we see an overall increase in activity in the second half of the month (could be that people waited their monthly salaries until around 15th). This may be a recurring condition, since on the last day of the month the number of clicks drops to the level witnessed at the beginning of the month. Finally, on every Saturday, activity languished a bit compared to the rest of the week (perhaps people typically sit in front of their computers less on Saturdays). These effects are intriguing and deserve deeper analysis with more data.

59

Persisting Results

The whole point of data preprocessing is to preserve the results for later use. The previous extension of `driver.py` only visualizes compressed data frames. Once the program is run, all that remains are diagrams and raw CSV files. It would be better to save data frames into the `results` folder, as depicted earlier in Figure 2-2. During later data analysis, these data frames could be read without starting everything all over. There are many available file formats. We are going to use Apache Parquet (see `https://parquet.apache.org`), which delivers a partitioned binary columnar storage format for data frames. The following are some of the benefits and important aspects regarding Parquet:

- It faithfully and efficiently writes and reads data frames.

- It allows sharing of data across a multitude of programming languages and frameworks.

- It nicely interoperates with the Hadoop ecosystem.

- Pandas has a very intuitive API for handling serialization and deserialization of data frames into/ from this format.

Listing 2-6 shows the final section of `driver.py` to save each data frame into a separate file (the part from Listing 2-5 should be commented out; see Exercise 2-2). A set of these files provides a clear demarcation line between this stage and the next one in a data science process.

Listing 2-6. Code to Write Data Frames into Parquet Files Using the `nyt_summary_<day number>.parquet` File Naming Convention

```
def save_stats(df, file_num):
    targetFile = 'nyt_summary_' + str(file_num + 1) +
    '.parquet'
```

```
df.columns = ['_'.join(column).rstrip('_') for column in
df.columns.values]
df.to_parquet('results/' + targetFile)
```

```
traverse('raw_data', save_stats)
print('Raw data files are successfully processed.')
```

It is mandatory to have string column names in a DataFrame instance before writing it into a Parquet file. The code shows how to convert a multilevel index into a flat structure, where the top index's name is prepended to its subordinate columns. This transformation is reversible, so nothing is lost.

After all files are written, you may browse them in the File Explorer. They are all the same size of 6KB. Compare this to initial CSV file sizes.

Parquet Engines

We didn't designate a specific Parquet engine in Listing 2-6, which means it was set to auto. In this case, Pandas tries to load pyarrow or fastparquet to handle read/write operations. If you don't have these installed, then you will receive the following error message:

```
ImportError: Unable to find a usable engine; tried using:
'pyarrow', 'fastparquet'.
pyarrow or fastparquet is required for parquet support
```

An easy to way to resolve this is to issue conda install pyarrow (you may also want to specify an environment). This command installs the pyarrow package, which is a Python binding for Apache Arrow (see http://arrow.apache.org). Arrow is a cross-language development platform for columnar in-memory format for flat and hierarchical data. With pyarrow it is easy to read Apache Parquet files into Arrow structures. The capability to effortlessly convert from one format into another is an important criterion when choosing a specific technology.

Restructuring Code to Cope with Large CSV Files

Breaking up data into smaller files, as we have done so far, is definitely a good strategy to deal with a massive volume of data. Nonetheless, sometimes even a single CSV file is too large to fit into memory. Of course, you can try vertical scaling, by finding a beefed-up machine, but this approach is severely limited and costly. A better way is to process files in *chunked* fashion. A *chunk* is a piece of data from a file; once you are done with the current chunk, you can throw it away and get a new one. This tactic can save memory tremendously but may increase processing time.

Listing 2-7 shows the restructured version of the summarize function that handles data chunk-by-chunk. There are lots of further improvements that may be done regarding how chunks are processed. However, the intention here is to convey the basic idea in a straightforward manner. The key point is that only when you're absolutely convinced that neither file sharding nor chunking helps should you switch to more advanced frameworks and setup. This is consistent with the golden tenet of data science to assume a simple resolution unless proven otherwise.

Listing 2-7. Restructured nyt_data_chunked.py to Showcase Chunked Data Processing

```
def summarize(data_file, chunksize):
    def q25(x):
        return x.quantile(0.25)

    def q75(x):
        return x.quantile(0.75)

    # Read and parse the CSV data file chunk-by-chunk.
    nyt_data = pd.DataFrame()
```

```
for chunk_df in pd.read_csv(
        data_file,
        dtype={'Gender': 'category'},
        chunksize=chunksize):

    # Segment users into age groups.
    chunk_df['Age_Group'] = pd.cut(
            chunk_df['Age'],
            bins=[-1, 0, 17, 24, 34, 44, 54, 64, 120],
            labels=["Unknown",
                    "1-17",
                    "18-24",
                    "25-34",
                    "35-44",
                    "45-54",
                    "55-64",
                    "65+"])

    # Create the click-through rate feature.
    chunk_df['CTR'] = chunk_df['Clicks'] /
    chunk_df['Impressions']
    chunk_df.dropna(inplace=True)
    chunk_df.drop((chunk_df['Clicks'] >
    chunk_df['Impressions']).nonzero()[0],
                inplace=True)

    # Append chunk to the main data frame.
    nyt_data = nyt_data.append(
            chunk_df[['Age_Group', 'Gender', 'Clicks', 'CTR']],
            ignore_index=True)
```

```
# Make final description of data.
compressed_nyt_data = \
    nyt_data.groupby(by=['Age_Group', 'Gender'])[['CTR',
    'Clicks']] \
            .agg([np.mean, np.std, np.max, q25, np.median,
            q75, np.sum])
return compressed_nyt_data
```

The traverse function accepts chunksize as a parameter, which is set by default to 10000. The driver.py script is redirected to use the new module, and all the rest runs unchanged.

Public Data Sources

Data analysis revolves around data, of course, so finding useful data sources is crucial for development and testing. In the absence of freely available data, you may synthetically generate some or rely on accrued corporate data. However, that usually isn't necessary, because many public data sources are available; most are constantly inspected and improved by community members, and frequently accompanied with descriptions about their intended use. There are also high-quality datasets embedded in popular data science frameworks. The following snippet lists what is shipped with scikit-learn (only some lines are shown, for brevity):

```
>> import sklearn.datasets as datasets
>> [item for item in dir(datasets) if not item.startswith('_')]
['base',
 'california_housing',
 'clear_data_home',
 'covtype',
 'dump_svmlight_file',
 'fetch_20newsgroups',
```

```
'fetch_20newsgroups_vectorized',
'fetch_california_housing',
'fetch_covtype',
'fetch_kddcup99',
'fetch_lfw_pairs',
'fetch_lfw_people',
...]
```

>> datasets.fetch_20newsgroups?
Signature: datasets.fetch_20newsgroups(data_home=None,
subset='train', categories=None, shuffle=True, random_state=42,
remove=(), download_if_missing=True)
Docstring:
Load the filenames and data from the 20 newsgroups dataset
(classification).

Download it if necessary.

```
=================   ==========
Classes                     20
Samples total            18846
Dimensionality               1
Features                  text
=================   ==========
```
Read more in the :ref:`User Guide <20newsgroups_dataset>`.

Parameters
----
data_home : optional, default: None
    Specify a download and cache folder for the datasets.
    If None,
    all scikit-learn data is stored in '~/scikit_learn_data'
    subfolders.

...

The following list is just a taste of public data sources accessible on the Internet and what types of access are offered:

- Data.gov (`https://www.data.gov`) is the U.S. federal government's data hub. In addition to loads of datasets, you can find tools and resources to conduct research and create data products.

- Geocoding API (`https://developers.google.com/maps/documentation/geocoding/start`) is a Google REST service that provides geocoding and reverse geocoding of addresses. This isn't a dataset per se but is important in many data engineering endeavors.

- GitHub repository (`https://github.com/awesomedata/awesome-public-datasets`) offers a categorized list of publicly available datasets. This is an example of a source aggregator that alleviates the burden of targeted searches for datasets.

- HealthData.gov (`https://healthdata.gov`) offers all sorts of health data for research institutions and corporations.

- Kaggle (`https://www.kaggle.com`) is a central place for data science projects. It regularly organizes data science competitions with excellent prizes. Many datasets are accompanied by kernels, which are notebooks containing examples of working with the matching datasets.

- National Digital Forecast Database REST Web Service (`https://graphical.weather.gov/xml/rest.php`) delivers data from the National Weather Service's National Digital Forecast Database in XML format.

- Natural Earth (`http://www.naturalearthdata.com`) is a public domain map dataset that may be integrated with Geographic Information System software to create dynamic map-enabled queries and visualizations.

- Open Data Network (see `https://www.opendatanetwork.com`) is another good example of a source aggregator. It partitions data based on categories and geographical regions.

- OpenStreetMap (`https://wiki.openstreetmap.org`) provides free geographic data for the world.

- Quandl (`https://www.quandl.com`) is foremost a commercial platform for financial data but also includes many free datasets. The data retrieval is via a nice API with a Python binding.

- Scikit Data Access (`https://github.com/MITHaystack/scikit-dataaccess`) a curated data pipeline and set of data interfaces for Python. It prescribes best practices to build your data infrastructure.

- Statistics Netherlands opendata API client for Python (`https://cbsodata.readthedocs.io`) is a superb example of using the Open Data Protocol (see `https://www.odata.org`) to standardize interaction with a source.

- UCI Machine Learning Repository (`http://archive.ics.uci.edu/ml/index.php`) maintains over 400 datasets for different machine learning tasks. Each set is properly marked to identify which problems it applies to. This is a classical example of a source where you need to download the particular dataset.

EXERCISE 2-1. ENHANCE REUSABILITY

The `driver.py` script contains lots of hard-coded values. This seriously hinders its reusability, since it assumes more than necessary. Refactor the code to use an external configuration file, which would be loaded at startup.

Don't forget that the current driver code assumes the existence of the input and output folders. This assumption no longer holds with a configuration file. So, you will need to check that these folders exist and, if not, create them on demand.

EXERCISE 2-2. AVOID NONRUNNABLE CODE

The `driver.py` script contains a huge chunk of commented-out code. This chunk isn't executed unless the code is uncommented again. Such passive code tends to diverge from the rest of the source base, since it is rarely maintained and tested. Furthermore, lots of such commented sections impede readability.

Restructure the program to avoid relying on comments to control what section will run. There are many viable alternatives to choose from. For example, try to introduce a command-line argument to select visualization, output file generation, or both. You may want to look at `https://docs.python.org/3/library/argparse.html`.

EXERCISE 2-3. AUGMENT DATA

Sometimes you must augment data with missing features, and you should be vigilant to search for available solutions. If they are written in different languages, then you need to find a way to make them interoperate with Python. Suppose you have a dataset with personal names, but you are lacking an additional `gender` column. You have found a potential candidate Perl module, `Text::GenderFromName` (see `https://metacpan.org/pod/`

`Text::GenderFromName`). Figure out how to call this module from Python to add that extra attribute (you will find plenty of advice by searching on Google).

This type of investigation is also a job of data engineers. People focusing on analytics don't have time or expertise to deal with peculiar data wrangling.

Summary

Data engineering is an enabler for data analysis. It requires deep engineering skills to cope with various raw data formats, communication protocols, data storage systems, exploratory visualization techniques, and data transformations. It is hard to exemplify all possible scenarios.

The e-commerce customer segmentation case study demonstrates mathematical statistics in action to compress the input data into a set of descriptive properties. These faithfully preserve main behavioral traits of users whose actions were preserved in raw data. The case study ends by examining large datasets and how to process them chunk-by-chunk. This is a powerful technique that may solve the problem without reaching to more advanced and distributed approaches.

In most cases, when dealing with common data sources, you have a standardized solution. For example, to handle disparate relational database management systems in a unified fashion, you can rely on the Python Database API (see `https://www.python.org/dev/peps/pep-0249`). By contrast, when you need to squeeze out information from unusual data sources (like low-level devices), then you must be creative and vigilant. Always look for opportunity to reuse before crafting your own answer. In the case of low-level devices, the superb library PyVISA (see `https://pyvisa.readthedocs.io`) enables you to control all kinds of measurement devices independently of the interface (e.g., GPIB, RS232, USB, and Ethernet).

Data-intensive systems must be secured. It isn't enough to connect the pieces and assume that under normal conditions all will properly work. Sensitive data must be anonymized before being made available for further processing (especially before being openly published) and must be encrypted both at rest and in transit. It is also imperative to ensure consistency and integrity of data. Analysis of maliciously tampered data may have harmful consequences (see reference [4] for a good example of this).

There is a surge to realize the concept of *data as commodity*, like cloud computing did with computational resources. One such initiative is coming from Data Rivers (consult [5]), whose system performs data collection, cleansing (including wiping out personally identifiable information), and validation. It uses Kafka as a messaging hub with Confluent's schema registry to handle a humungous number of messages.

Another interesting notion is data science as a service (watch the webinar at [6]) that operationalizes data science and makes all necessary infrastructure instantly available to corporations.

References

1. Cathy O'Neil and Rachel Schutt, *Doing Data Science*, O'Reilly, 2013.

2. "Using Segmentation to Improve Click Rate and Increase Sales," Mailchimp, `https://mailchimp.com/resources/using-segmentation-to-improve-click-rate-and-increase-sales`, Feb. 28, 2017.

3. Wes McKinney, *Python for Data Analysis: Data Wrangling with Pandas, NumPy, and IPython, 2nd Edition*, O'Reilly, 2017.

4. UC San Diego Jacobs School of Engineering, "How
    Unsecured, Obsolete Medical Record Systems and
    Medical Devices Put Patient Lives at Risk," `http://`
    `jacobsschool.ucsd.edu/news/news_releases/`
    `release.sfe?id=2619`, Aug. 28, 2018.

5. Stephanie Kanowitz, "Open Data Grows
    Up," *GCN Magazine*, `https://gcn.com/`
    `articles/2018/08/28/pittsburgh-data-rivers.`
    `aspx`, Aug. 28, 2018,.

6. "Introducing the Data Science Sandbox as a
    Service," Cloudera webinar recorded Aug. 30,
    2018, available at `https://www.cloudera.com/`
    `resources/resources-library.html`.

CHAPTER 3

Software Engineering

An integral part of data science is executing efficient computations on large datasets. Such computations are driven by computer programs, and as problems increase in size and complexity, the accompanying software solutions tend to become larger and more intricate, too. Such software is built by organizations structured around teams. Data science is also a team effort, so effective communication and collaboration is extremely important in both software engineering and data science. Software developed by one team must be comprehensible to other teams to foster use, reuse, and evolution. This is where software maintainability, as a pertinent quality attribute, kicks in. This chapter presents important lessons from the realm of software engineering in the context of data science. The aim is to educate data science practitioners how to craft evolvable programs and increase productivity.

This chapter illustrates what it means to properly fix existing code, how to craft proper application programming interfaces (APIs), why being explicit helps to avoid dangerous defects in production, etc. The material revolves around code excerpts with short, theoretical explanations. The focus is on pragmatic approaches to achieve high quality and reduce wasted time on correcting bugs. This will be an opportunity for data product developers of all skill levels to see rules in action, while managers will get crucial insights into why investing in quality up front is the right way to go. You will learn how and why principles of software engineering in the context of *The Zen of Python* materialize in practice.

© Ervin Varga 2019
E. Varga, *Practical Data Science with Python 3*,
https://doi.org/10.1007/978-1-4842-4859-1_3

According to a 2018 article describing a survey by Stripe (see reference [1] in the "References" section at the end of the chapter), "Companies around the world are collectively losing about $300 billion a year by wasting developer resources on fixing maintenance issues..." This definitely impacts companies that specialize in data science–oriented, large-scale Python software systems, too. The following statement nicely emphasizes this truism:

After all, the critical programming concerns of software engineering and artificial intelligence tend to coalesce as the systems under investigation become larger.

—Alan J. Perlis, Foreword to *Structure and Interpretation of Computer Programs, Second Edition* (MIT Press, 1996)

Usually, data science is described as a discipline embracing mathematical skills, computer science skills, and domain expertise. What is lacking here is the notion of engineering and economics (science of choices). Software engineering aspects are important to create maintainable, scalable, and dependable solutions. ISO/IEC/IEEE Systems and Software Engineering Vocabulary (SEVOCAB; see `https://pascal.computer.org`) defines *software engineering* as "the application of a systematic, disciplined, quantifiable approach to the development, operation, and maintenance of software; that is, the application of engineering to software." This chapter tries to cover the gap and explain the role of engineering and economics in data science.

Changes programming languages, required to address the challenges of large-scale software development, are imminent; shifts in Python are already underway. In this respect, Python will share the destiny of many other enterprise-level languages (it is instructive to evoke Java's evolution over the last two decades). This chapter mentions some crucial quality assurance tools that augment Python's expressiveness to be a better fit for enterprise software systems.

At any rate, there is no point in recounting known facts or solely musing about theory. Therefore, the content of this chapter is rather atypical and centers on hidden but crucial aspects of the software industry.

Characteristics of a Large-Scale Software System

The biggest difference between a large-scale software system and a small-scale one revolves around the software's life cycle model. It refers to all the phases of a software product, from initial development, through exploitation, maintenance, evolution, and retirement. To make the discussion simple, our notion of *large-scale* is a software system that cannot be implemented by a small team; is part of a complex, socio-technical, software-intensive system; and whose retirement would have a huge business impact. Software maintenance and evolution is usually the longest phase in a large-scale software system's life cycle; for some safety-critical software systems, it may span a couple of decades. This means that most activities will happen after the deployment of the first production version.

It is instructive to use a tournament chess game as an analogy to the software life cycle model. As with software, such a chess game is preceded by an initial preparation and planning phase (e.g., analyze your opponent's style of play and opening repertoire, recall past experience with the opponent, compare your current standings, decide whether you want to aggressively push for a win, etc.). You cannot approach the game as amateurs do by simply making moves based on the current position of the chess pieces. During the game you balance resources (material, time, position, etc.), apply theory, use experience, and progress through stages (opening, middle game, and end game). The stages of a chess game are mapped to the phases of the software life cycle model in Figure 3-1.

To keep the cycle rolling, a holistic approach to software engineering is required. You must evaluate a broad spectrum of possibilities to properly judge the potential vectors of changes. These dictate where your software must be flexible in order to accommodate future additions and improvements. In each round, you must also address waves of change requests; this is similar to chess, where you must constantly monitor what your opponent is doing and alter your strategy accordingly. Finally, as in chess, if you mess up the requirements or devise a wrong architecture, then it is very difficult to make corrections later. In chess, a bad opening would turn your game into a struggle for a draw.

While in chess everything ends in the last stage, this is completely different in the development of a large-scale software system. The cycles are chained and encompass developments of new features. There is even a view that software maintenance and evolution is a continuation of the development phase, and in this respect, we talk about perpetual evolution (successful improvements and extensions of a product). In some sense, developing a small software product is like a single chess game, while developing a large-scale software system is akin to a very long tournament.

Besides adding new capabilities to existing software, there is also a need for maintenance. Fixing bugs is a prime example of this activity. For economical maintenance and evolution, the software must be *maintainable*. According to the ISO/IEC 25010 standard, maintainability as a quality attribute (testability is its subattribute) represents the capability to efficiently incorporate changes into the code base. According to IEEE 14764, we can differentiate between the following four categories of maintenance, which may be classified as either of two major types (correction or enhancement):

- **Corrective maintenance (correction)**: Related to fixing failures (observable problems causing some harm in production) that are usually reported by a customer.

- **Adaptive maintenance (enhancement):** Pertains to modifications due to changes in the environment (e.g., business process, external regulatory law, platform, technology, etc.). The goal is to proactively keep the software usable.

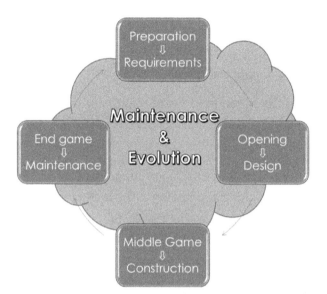

Figure 3-1. *Mapping of chess stages to software's life cycle model*

- **Preventive maintenance (correction):** Related to fixing faults before they transform into a failure. Suppose that your code has a function accepting a numeric argument that is used as a divisor. Without checking against zero, it may generate an exception. However, if that argument will never receive a value of zero, then this potential fault cannot become a failure (operational fault). This requires a thorough understanding of how a customer will use a product, as some faults may never turn into failures under some circumstances.

- **Perfective maintenance (enhancement)**: Deals
  with all sorts of proactive changes to boost the
  quality of the software (including maintainability).
  This activity is essential to combat the detrimental
  effects of continuous changes to a code base during
  maintenance and evolution. As new features are added,
  complexity increases, and the existing code structure
  deteriorates. This is nicely documented in *Lehman's
  laws of software evolution*.

All the previously mentioned maintenance activities and categories
must be evaluated against the type of the target software system. There are
three such types:

- **S type system**: Has a finite enumerable problem space
  with a fixed environment. Once the initial requirements
  are specified, nothing will change in the future (here,
  we don't count potential user interface enhancement
  as impactful changes). For example, the popular
  Tic-Tac-Toe game is an S type system.

- **P type system**: Has a complex problem space with a
  fixed environment. Various heuristics are applied to
  search the problem space. An improvement to the
  abstract model does induce a considerable change in
  the system. For example, popular board games like
  chess and Go are P type systems.

- **E type system**: The most complex system type due to a
  highly dynamic environment. It is impossible to predict
  with 100% accuracy how the environment will fluctuate
  in the future. Unfortunately, nobody can assure that a
  new law will not disturb the original architecture of the
  system. A perfect recent example of this is enactment

of the EU's General Data Protection Regulation (GDPR; see `https://gdpr-info.eu`), compliance with which required many organizations to make considerable investments in software system updates. All in all, large-scale data science products belong to this category.

The total cost of ownership (TCO) of a software system is composed of the following elements:

- The initial development of a new product

- Enhancement maintenance-type activities

- Correction maintenance-type activities

- Customer support

Evidently, the ratio is 3:1 between maintenance and initial development. Consequently, we should beware of being fast at the expense of quality. Many immature companies nurture a false belief that tooling is the silver bullet. They will both speed things up and compensate for human deficiencies (lack of knowledge and experience, inattentiveness, etc.). As you will later see, the only solution is to invest in people before anything else.

Software Engineering Knowledge Areas

Software engineering is composed of 15 primary knowledge areas (see also [3] in the "References" section):

- Software Requirements

- Software Design

- Software Construction

- Software Testing

- Software Maintenance

- Software Configuration Management
- Software Engineering Management
- Software Engineering Process
- Software Engineering Models and Methods
- Software Quality
- Software Engineering Professional Practice
- Software Engineering Economics
- Computing Foundations
- Mathematical Foundations
- Engineering Foundations

To fully specify the scope of software engineering, we also need to include the following seven related disciplines:

- Computer Engineering
- Computer Science
- General Management
- Mathematics
- Project Management
- Quality Management
- Systems Engineering

Computer science and mathematics are obvious foundations, but people have difficulties grasping systems engineering (read the sidebar "Success or Not?"). The latter is a much broader discipline that gives meaning to software products by putting them into a socio-technical context.

SUCCESS OR NOT?

I am pretty sure that you've received many auto-generated birthday greeting e-mail messages up so far. Suppose that the software generating these is impeccable. It searches the user database each day and filters out all users who were born on that date. It then produces messages using a common template and sends them out. Are you happy when you receive such a message? Personally, I think they are extremely annoying. You likely are glad to get a message from your spouse, kid, relative, friend, or colleague, but probably not from a machine programmed to mechanically deliver greetings.

From the viewpoint of software engineering, the company may say that this software is a total success. Nonetheless, if no one wants to receive auto-generated greetings, it is a complete failure from the standpoint of systems engineering. The software is useless despite working perfectly.

Rules, Principles, Conventions, and Standards

All the terms in the title of this section are interrelated and differ mostly in their level of rigor. Conventions are agreements between parties, while standards are formalized specifications that are managed by appropriate standardization bodies (for example, *IEEE 754-2019 - IEEE Standard for Floating-Point Arithmetic*; see https://standards.ieee. org/standard/754-2008.html). Unlike physics, the software engineering discipline usually doesn't have laws, although exceptions exist. Rules and principles are in the middle of the spectrum between conventions and standards. All of them are intended to impose constraints on system design to boost interoperability, integration, and productivity. Maybe it sounds

weird, but judicious limitation of choices actually channels creativity; the
objective is to avoid reinventing the wheel all the time. Furthermore, rules
and principles synthesize knowledge of a field in a straightforward manner
that helps attain uniformity and quality. Python is also bundled with a set
of rules called *The Zen of Python*, as shown in Figure 3-2. The standard
quality model is attached to the list to remind you how these items are
important.

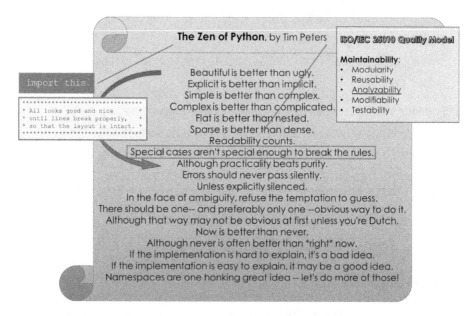

Figure 3-2. *The set of rules that you get after executing* `import this`
in a Python console

As you can see, according to The Zen of Python, it is crucial to eschew
the temptation of treating rules and principles as dogmas. This is superbly
illustrated by contrasting the first and ninth rules (which I have connected
with a curved arrow). So, even The Zen of Python warns us about this
problem. My commentary inside the box on the left provides an example
of why vehemently trying to satisfy a rule could cause problems. On the
other hand, there is an important meta-rule (the eighth rule, shown with
a border) to remind us that it is equally wrong to blatantly ignore rules.

You must know what you're doing before departing from any sort of edict. "Readability counts" has a central location, with good reason. Software engineers spend most of their time during maintenance and evolution on understanding code.

This chapter implicitly follows the rules listed in Figure 3-2; the exposition of topics relies on simple examples (the Zen mentions implementation details instead of narrative). An absence of such examples could indicate a vague topic, an inept lecturer, or both. The illustrations that follow are indeed as simple as possible to keep your focus on the core subject. There is a key Agile principle (see `https://agilemanifesto.org/principles`) about simplicity that states "Simplicity--the art of maximizing the amount of work not done--is essential." Ironically, this is the most misapplied tenet. The work required to realize what should be expelled is everything but simple.

Software engineers leverage quality assurance tools to effortlessly apply rules (they can also enforce rules, especially in combination with a build server). The Python ecosystem is abundant with such tools. The following list is just a sample of what is available:

- **Pylint** (`https://www.pylint.org`) offers features such as:

    - Code style checking (visit `https://www.python.org/dev/peps/pep-0008`)

    - Reverse engineering support (integrated with Pyreverse to produce UML diagrams from code)

    - Refactoring help (such as detection of duplicated code)

    - IDE integration (for example, it is an integral part of Spyder)

    - Error detection (customizable checkers)

- **Codacy** (https://www.codacy.com) provides automated code reviews.

- **CodeFactor** (https://www.codefactor.io) also provides automated code reviews.

- **Radon** (https://pypi.org/project/radon) offers features such as:

  - McCabe's cyclomatic complexity (explained a bit later in the chapter)

  - Raw metrics, including source lines of code (SLOC), comment lines, blank lines, etc.

  - Halstead metrics

  - Maintainability index

- **Mypy** (https://mypy-lang.org) is an experimental, optional static type checker for Python. Statically typed languages are known to have advantages over dynamically typed ones for large code bases. The ability of a compiler to detect type-related omissions early is often a lifesaver. So, it isn't surprising to see this tool on the plate.

The preceding tools calculate quantitative proxies for qualitative aspects of a code base. Consequently, they can only report and act upon efficiently measurable indicators. Often, the most important traits of a software development endeavor are hard or impossible to measure. People-related elements belong to this class. The next section explains this phenomenon in more detail.

Context Awareness and Communicative Abilities

In general, we may discern three major personality types among software engineers[1]:

Context and knowledge oblivious

The recipient will blindly follow instructions without seeking out help for further clarification. This type of person will guess instead of basing his decision on facts. Moreover, due to lack of knowledge, he will make up ad hoc rules/principles, as he judges appropriate. You may try excessive micromanagement, which is equally troublesome for both parties. This behavior is depicted in Figure 3-3.

Knowledgeable

This class is very similar to the previous one, the only difference being that this type of person has proper knowledge and thus will stick to major rules and principles of the discipline. This behavior is depicted in Figure 3-4.

Context aware and knowledgeable

This type of person seeks to maintain a constant feedback loop by proactively asking questions, as needed. This behavior is depicted in Figure 3-5.

---

[1]You may also want to read the blog at `https://people.neilon.software`. This article depicts in a funny way various stereotypes in relation to roles. Managing people is tedious, and social life in an organization is as important as technical expertise.

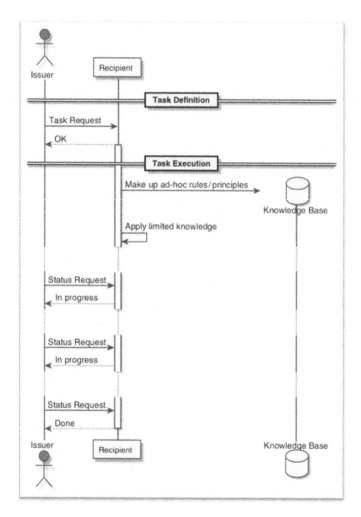

Figure 3-3. *The final "Done" message doesn't imply that the business goals are met (adapted with permission from reference [2]). The quality of the result is also questionable.*

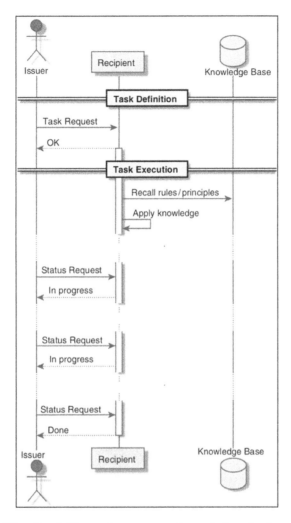

Figure 3-4. *The final "Done" message doesn't imply that the business goals are met (adapted with permission from [2]). The quality of the result may be proper.*

Software engineers must be conscious of the language of business and talk to management in terms of money flows. Otherwise, there is a danger of destroying the precious feedback loop between development and management, resulting in two isolated factions (something that frequently happens in organizations). Such separation has detrimental consequences on success of the business.

Asking the proper question (the question itself is the treasure on the island in Figure 3-5), properly formulating the problem, and expressing choices in the language of business are most important to keep the valuable feedback loop active. There are two preconditions for attaining quality via feedback loops: having time to ask questions and create alternative designs, and knowing how to talk to business (management). Both conditions are usually neglected by software professionals.

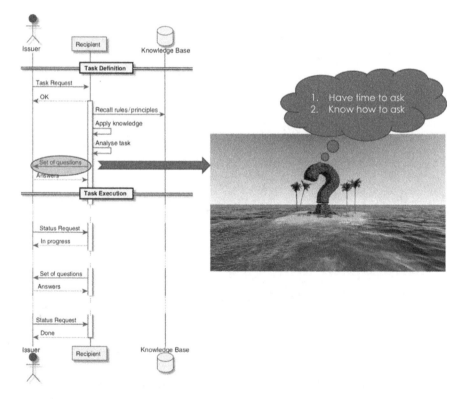

Figure 3-5. *This person tries to solve the right problem with high quality (adapted with permission from [2])*

Reducing Cyclomatic Complexity

Cyclomatic complexity is a measure that increases with the number of alternative execution flows of a program. For example, every conditional statement bumps the score. A less complex code will have a lower cyclomatic number. Suppose that management has decided to reduce the cyclomatic complexity of a code base by allowing developers to look for opportunities to simplify the current code.

Soon, a *context and knowledge oblivious* developer stumbles across the function that sorts an array, as shown in Listing 3-1 (see sort_original. py in the cyclomatic_complexity folder). For now, assume that this functionality is unique and not available as a built-in function.

Listing 3-1. Function to Sort an Input Array into Ascending Order

```
def sort(data):
    for i in range(len(data)):
        for j in range(len(data)):
            if data[i] > data[j]:
                data[i], data[j] = (data[j], data[i])
    return data
```

After executing sort([-2, -4, 100, 3, 4.1]) the function returns [100, 4.1, 3, -2, -4]. Sure, it will also alter the input array, as it sorts in place.

Our eager developer also runs static analysis in Spyder (essentially, runs the Pylint tool). Here is the report:

```
********* Module sort_original
C0304:  8,0: : Final newline missing
C0111:  1,0: : Missing module docstring
C0111:  3,0: sort: Missing function docstring
C0200:  4,4: sort: Consider using enumerate instead of
iterating with range and len
```

```
CO200:  5,8: sort: Consider using enumerate instead of
iterating with range and len
```

```
Your code has been rated at 1.67/10
```

After some time, he discovered a "brilliant" way to reduce the cyclomatic complexity by eliminating the inner if statement (the score would surely drop). Listing 3-2 shows the "improved" version of the sorting routine.

Listing 3-2. Version of Sort Function with a Lower Cyclomatic Complexity

```
def sort(data):
    for i in range(len(data)):
        for j in range(len(data)):
            avg = (data[i] + data[j]) / 2.0
            diff = abs(data[i] - avg)
            data[i] = avg - diff
            data[j] = avg + diff
    return data
```

After executing sort([-2, -4, 100, 3, 4.1]) the function now returns [100.0, 4.100000000000001, 3.0, -2.0, -4.0] (so, it doesn't quite work, but is "close" enough). The static analysis report was even more compelling support for the developer's conclusion that this was a great achievement (he was astute to also fix the previously reported newline ending issue):

```
********* Module sort_new
C0111:  1,0: : Missing module docstring
C0111:  3,0: sort: Missing function docstring
CO200:  4,4: sort: Consider using enumerate instead of
iterating with range and len
```

```
CO200:  5,8: sort: Consider using enumerate instead of
iterating with range and len
```

```
------------------------------------------------------------------
Your code has been rated at 5.00/10
```

All in all, besides reducing the cyclomatic complexity, the global evaluation score went up from 1.67 to an astonishingly high 5! What has happened? Has Pylint failed? Obviously, the end result is far from what was expected by management. The code doesn't even function properly anymore.

Caution   When basic mathematical logic fails, then nothing else matters. In this case, the developer believed that implication is equivalent to its converse. This is false; that is, $\varphi \Rightarrow \psi \nLeftrightarrow \psi \Rightarrow \varphi$! A lower cyclomatic complexity doesn't imply a less complex code. The corollary is that tools alone are not enough! People should drive the tools, and not vice versa.

Cone of Uncertainty and Having Time to Ask

The *Cone of Uncertainty* represents a graphical model about best-case reduction of estimation error during the lifetime of a project (see Exercise 3-1 and visit https://www.construx.com/software-thought-leadership/books/the-cone-of-uncertainty). The Cone of Uncertainty tells us that in a well-managed project, we may expect to have less variability in project scope (size, cost, or features) as the project progresses toward its end. Less variability implies higher predictability and accuracy of estimates. At the very beginning of a project, the variability is huge (in the range of [0.25×, 4×]). Anything <1× signifies underestimation, while >1× signifies overestimation. Obviously, 1× represents total predictability.

The Cone of Uncertainty is biased toward the lower bound; in other words, there is a higher tendency to underestimate.

The idea is to eliminate critical reasons for this unpredictability first, thus narrowing the cone as rapidly as possible. The following events are major milestones in diminishing variability (we start the list with an initial project vision already formulated):

- Initial concept

- Approved product definition

- Marketing requirements complete

- Detailed technical requirements complete and user interface design complete (expected variability is around [0.8×, 1.25×])

- Detailed design complete

Both project leaders and software engineers must differentiate between *estimation* and *commitment*. The latter is only allowed when the cone is sufficiently narrow (when the variability is in the range of [0.8×, 1.25×]). Making commitments earlier just depletes your time to focus on quality, as unrealistic expectations cause stress and hasty work. Moreover, all participants in the project must work hard to keep the cone narrowing. Otherwise, it will look more like a cloud, with equally high variability throughout the project. The cone doesn't narrow by itself over time.

There are three possible scenarios, depending on when commitments are made:

- **Danger zone:** Variability in the range [1.25×, 4×] with linear penalty (per Parkinson's law, which states that people will always fill in their time slots with some work): overestimation→undercommittment→low pressure→moderate quality

- **Danger zone**: Variability in the range [0.25×, 0.8×] with exponential penalty:
  underestimation➝overcommittment➝extreme pressure➝low quality

- **Allowed zone**: Variability in the range [0.8×, 1.25×] without penalty:
  good estimation➝good committment➝optimal pressure➝high quality

The Cone of Uncertainty and the heuristics around estimation and commitment are as important to data science projects as they are to any other software projects. Both are financed by stakeholders who make decisions based upon project plans.

Fixing a Bug and Knowing How to Ask

This section presents a series of scenarios pertaining to fixing a bug in a corporate setting. Assume for these scenarios that the company follows practices dictated by modern Agile methods. More specifically, any reported issue progresses through the following states:

1. An issue is discovered and reported in a bug tracking system.

2. Work is estimated.

3. The issue is assigned and prioritized.

4. Proper unit tests are written.

5. The bug is fixed, and all tests are passing.

6. The status is set and a hot fix is deployed.

Assume that the first three items have been accomplished already, so we are left with the remaining three items. The bug is reported for the code

as shown in Listing 3-3 (saved under double_preceding1.py in the bug_
fixing folder). The code is not beautiful, and the implementation doesn't
reflect the intention highlighted in the embedded documentation.

Listing 3-3. Code with a Defect As Reported by Customer

```
from array import array

def double_preceding(x: array) -> None:
    """Transforms the array by setting x[i] = 2 * x[i-1] and
    x[0] = 0.

    >>> x = array('i', [5, 10, 15])
    >>> double_preceding(x)
    >>> x
    array('i', [0, 10, 20])
    """

    if x:
        temp = x[0]; x[0] = 0
        for i in range(1, len(x)):
            x[i] = 2 * temp; temp = x[i]
```

The first task is to write unit tests that include the one that exposes the
bug, as shown in Listing 3-4 (saved under test_double_preceding.py).
This is a good practice to avoid later regressions.

Listing 3-4. Unit Test for the Function from Listing 3-3

```
from array import array
import unittest
from double_preceding1 import double_preceding
```

```
class TestDoublePreceding(unittest.TestCase):
    """Tests for double_preceding function."""

    def test_already_arranged(self):
        """Test with already arranged values."""
        argument = array('i', [5, 10, 15])
        expected = array('i', [0, 10, 20])
        double_preceding(argument)
        self.assertEqual(expected, argument)

    def test_identical(self):
        """Test with multiple identical values."""
        argument = array('i', [0, 1, 1])
        expected = array('i', [0, 0, 2])
        double_preceding(argument)
        self.assertEqual(expected, argument)

    def test_empty(self):
        """Test with an empty array."""
        argument = []
        expected = []
        double_preceding(argument)
        self.assertEqual(expected, argument)

if __name__ == "__main__":
    unittest.main()
```

Running this test suite produces the following error report, so the bug is reproduced:

```
..F
================================================================
FAIL: test_identical (__main__.TestDoublePreceding)
Test with multiple identical values.
```

```
Traceback (most recent call last):
  File "/Users/evarga/Projects/pdsp_book/src/ch3/bug_fixing/
  test_double_preceding.py", line 22, in test_identical
    self.assertEqual(expected, argument)
AssertionError: array('i', [0, 0, 2]) != array('i', [0, 0, 0])
```

```
Ran 3 tests in 0.009s
```

```
FAILED (failures=1)
```

Now, the job is to make a fix and rerun the tests to see if they all pass. It is easy to notice that the original code ignores all elements of an array except the first one. Listing 3-5 (saved under double_preceding2.py) shows the correct version.

Listing 3-5. "Correct" Version of the Code

```python
from array import array

def double_preceding(x: array) -> None:
    """Transforms the array by setting x[i] = 2 * x[i-1] and
    x[0] = 0.

    >>> x = array('i', [5, 10, 15])
    >>> double_preceding(x)
    >>> x
    array('i', [0, 10, 20])
    """

    if x:
        temp = x[0]; x[0] = 0
        for i in range(1, len(x)):
            temp_2x = 2 * temp; temp = x[i]; x[i] = temp_2x
```

All tests are passing now, so the developer had announced that the job is finished. At least, this is definitely the case from the viewpoint of tools and an established process in our fictional company. What do you think? The solution uses two ugly temporary variables and is a superb example of a simple but needlessly complicated code (look up the rule about this in The Zen of Python). In companies where only quantity matters (such as the number of fixes per some time period), this is the kind of maintenance work that happens. Over time the code base drifts into a territory of confusion, unreadability, and fragility.

A Better Fix

The developer doesn't want to follow the same path and decides to spend more time on also refactoring the code to improve maintainability. He comes up with the fix as shown in Listing 3-6 (saved under double_ preceding3.py).

Listing 3-6. An Improved Fix That Tries to Clean Up the Mess in the Original Code

```
from array import array

def double_preceding(x: array) -> None:
    """Transforms the array by setting x[i] = 2 * x[i-1] and
    x[0] = 0.

    >>> x = array('i', [5, 10, 15])
    >>> double_preceding(x)
    >>> x
    array('i', [0, 10, 20])
    """
```

```
if x:
    for i in range(-1, -len(x), -1):
        x[i] = 2 * x[i - 1]
    x[0] = 0
```

Observe that the code is literally following the formula from the embedded documentation. Sure, it also fixes the original bug. The developer has determined to step back and prevent being mentally locked into the current solution. The following quote from Albert Einstein buttresses this mindset: "No problem can be solved from the same level of consciousness that created it."

Nonetheless, there is a problem! The extra time equates to an outlay of money by the company, so management will want to know the reason behind such unexpected expenditure. This is where developers get trapped and demolish future opportunities for similar enhancements.

Scenario 1: The Developer Doesn't Speak the Language of Business

Typical arguments presented by a developer who doesn't speak the language of business are summarized below (he usually starts with the most exciting technical thing irrespective of the broader context):

- I have eliminated two temporary variables.

- The code is now beautiful and fully aligned with the embedded documentation.

- It is faster.

- It uses less memory.

- It has fewer SLOC.

- I have improved the global evaluation score from 4.44 up to 5.

Honestly, none of these are decipherable to managers; they hear them as technical mumbo-jumbo. Their reaction is typically along the lines of "The money was spent on NOTHING! Nobody approved you to work on this!"

Scenario 2: The Developer Does Speak the Language of Business

A developer who speaks the language of business knows software engineering economics, which is the science of choice for expressing options in terms of cash flows to ensure that they resonate with key business drivers. The arguments presented by this type of developer are entirely different from those listed in scenario 1:

- I have spent T_1 time to understand the original code.

- I have spent an additional T_2 time to refactor it to improve maintainability; no new risks were added.

- This code is central to the core functioning of our system.

- Other developers will only need to spend T_3 time to understand it, where T_3 is 20% of T_1.

- The $Return\ on\ Investment = \dfrac{Maintenance\ Savings}{Refactoring\ Cost} = \dfrac{0.8T_1}{T_1 + T_2}.$

This developer is absolutely aware of her role in the company and knows how the software that she has modified contributes to the business. She most likely carefully listens to what the CEO is talking about during quarterly company meetings to better align her work with the company's mission and strategic objectives. Furthermore, she understands that management wants to hear about risks, which ordinary developers typically don't realize, to their own detriment (they usually have poor skills in risk management).

The reaction of management in this scenario likely would be completely dissimilar to the previous case, with full approval given to the developer. This is how developers can get support for further improvements.

A More Advanced Fix

In this case, in addition to taking into account the oddity of the original code version, the developer has managed to boost performance. He has delivered the fix as shown in Listing 3-7 (saved under double_preceding4.py).

Listing 3-7. A Fast, Vectorized Version

```
import numpy as np

def double_preceding(x: np.ndarray) -> None:
    """Transforms the array by setting x[i] = 2 * x[i-1] and
    x[0] = 0.

    >>> x = np.array([5, 10, 15])
    >>> double_preceding(x)
    >>> x
    array([ 0, 10, 20])
    """

    if x.size != 0:
        x[:-x.size:-1] = 2 * x[-2::-1]
        x[0] = 0
```

Clearly, this variant is slightly more convoluted than the previous one. This shouldn't surprise you, as it is a known fact that performance optimization hinders maintainability. In data science projects, some performance-related decisions are considered as a norm, like using the NumPy framework from the very beginning.

Scenario 1: The Developer Doesn't Speak the Language of Business

The classical arguments in this case would be along the lines of the following:

- I have solved the problem with NumPy, which nowadays is used by everybody.

- I have eliminated the for loop and reduced SLOC.

- The code is much faster now due to vectorized operations.

- This function will serve as an example of NumPy for other developers.

Again, management will hear this as purely technical rhetoric. The concomitant reaction of management would likely be negative, such as "So, this madness won't stop here? Nobody is allowed to be NumPying around without authorization!"

Scenario 2: The Developer Does Speak the Language of Business

The proper business-associated arguments are listed here:

- After T time, I've improved the performance of a critical part of our product, based upon profiling. This has created a performance budget that didn't exist previously.

- The function's higher speed increases throughput in processing client requests. This reduces the risk of losing clients in peak periods and may positively impact the conversion rate.

- NumPy (a new framework) was introduced with minimal risks:

  - It is a popular open-source Python framework with a huge and vibrant community.

  - It is well maintained and documented.

  - I am an experienced user of this framework. No immediate extra costs (training time, consultants, etc.) are required.

This developer is acquainted with the architecture, so he knows where and why to improve the performance budget. Moreover, he takes care to explain how risks are mitigated with an introduction of a new framework. Developers quite often recklessly import packages just because they have seen neat stuff in them. If this trend is left without control in a company, then all sorts of problems may arise (not to mention pesky security issues). All in all, the preceding arguments again are likely to receive a green light from management.

Handling Legacy Code

A software system may transform into legacy code for many reasons: it uses an outdated technology; it is utterly non-evolvable due to its deteriorated quality; it is pushed out by novel business solutions; and so forth. Nevertheless, sometimes you will need to handle such code. The biggest conundrum is to reverse engineer it in an attempt to recover the necessary information to grasp its design and implementation. Don't forget that legacy code is sparsely covered by automated tests and isn't properly documented. All that you have is the bare source code. This section emphasizes the importance of documentation (test cases are a type of documentation) to avoid developing code that is essentially legacy code after its first release. The next chapter is devoted to JupyterLab, which is all about this matter.

Understanding Bug-Free Code

The source code is actually a snapshot that preserves the outcome of all design decisions in the moment when it was taken. How much is missing from such a snapshot is tightly related to the source code's quality. It is pure myth that you can recover everything even from low-quality source code. Suppose you stumble across the function shown in Listing 3-8 (saved under puzzle1.py in the puzzles folder).

Listing 3-8. Our First Bug-Free Legacy Code Example

```python
def puzzle1(n):
    p = 0; w = 1; s = n

    while w <= n:
        w <<= 2

    while w != 1:
        w >>= 2
        f = p + w
        p >>= 1

        if s >= f:
            p += w
            s -= f
    return p
```

One effective way to understand the function is to watch its behavior externally. Running it inside a loop produces the following output (the first column is the input, while the second one is the output):

0,0
1,1
2,1
3,1

4,2
5,2
...
8,2
9,3
10,3
...
15,3
16,4
17,4
...
23,4
24,4
25,5
26,5
...

It isn't hard to figure out what puzzle1 is doing. Ostensibly, it could be replaced with the following alternative:

```
def isqrt(n):
    from math import sqrt
    return int(sqrt(n))
```

We have fully recovered the behavior.[2] Is behavior all that matters? The original version doesn't rely on the math package and works solely with integers. Is this fact important? Well, these may be very important details. What if the code must be portable and capable of running on constrained devices that don't support floating point arithmetic? We need more information.

---

[2]We still don't understand the underlying algorithm. That is another level of reverse engineering. Apparently, we may control the depth of our quest.

Understanding Faulty Code

Now our job is much harder because there is a discrepancy between the code's current behavior and intended behavior. Remember that even this detail is hidden from us; that is, we don't know in advance whether code that is foreign to us is correct or not. All in all, we must reverse engineer both the actual and intended behavior. Listing 3-9 contains this code (saved under `puzzle2.py`).

Listing 3-9. Mysterious Code That Also Contains a Bug That We Need to Uncover

```
def puzzle2(bytes):
    f = [0] * 255
    s = k = 0

    for b in bytes:
        f[b] += 1

    s += f[k]
    k += 1
    while s < len(bytes) / 2:
        s += f[k]
        k += 1
    return k
```

The `bytes` parameter is a strangely named list, while `f` is some list of counters. Here are some sample runs (the bs denote the elements of `bytes`):

b1,	b2,	b3,	b4,	b5,	b6,	Output
2	2	3	3	3	3	**4**
2	2	2	2	3	3	3
2	2	2	4	4	4	**3**
1	2	2	2	3	14	3

It seems that the output is a median element, but the two output values shown in bold are clearly strange. Our next tactic is to try fixing the code by leveraging our hypothesis about the intended behavior (see also reference [5] about a similar puzzle in relation to APIs). Listing 3-10 shows the altered version (saved under puzzle2b.py).

Listing 3-10. Our Attempt to Come Up with a Bug-Free Version by Swapping the Bold Lines in Listing 3-9

```
def puzzle2(bytes):
    f = [0] * 255
    s = k = 0

    for b in bytes:
        f[b] += 1

    k += 1
    s += f[k]
    while s < len(bytes) / 2:
        k += 1
        s += f[k]
    return k
```

The new version generates the following output for the same input as before:

b1	b2	b3	b4	b5	b6	Output
2	2	3	3	3	3	3
2	2	2	2	3	3	2
2	2	2	4	4	4	**2**
1	2	2	2	3	14	2

All seems OK except the bold output value. Unfortunately, we cannot guess the correct answer between the following two choices:

1. The output is the nearest existing number k, such that the quantity of numbers less than k is closest to the quantity of numbers greater or equal to k.

2. The output is the nearest existing number k, such that the quantity of numbers less than or equal to k is closest to the quantity of numbers greater than k.

With faulty code we were not even able to fully recover the behavior.

The Importance of APIs

An API plays the following major roles in a system:

- Protects clients from internal system changes

- Protects the system against careless usage

- Allows management of technical debt

- Prevents entropy from flowing from one side to the other

An API must efficiently communicate the intended usage and behavior of some piece of a system. Furthermore, it should enable a client to gradually comprehend the target system, as needed. An API that forces you to understand everything before being able to interact with a system is a failure. Equally bad is to be forced to look into the source code to understand what a function or class performs. Structuring an API should revolve around the entropy model regarding the matching system.

Suppose you don't know anything about a system S. Therefore, your confusion and uncertainty are capped out. The level of entropy may be expressed as

$$H(S) = -\sum p(S = s_i) \log p(S = s_i)$$

(1)

where s_i is an assumed state of the system.

The API is a set of artifacts (usually abstractions); in other words, $API = \{a_1, a_2, ..., a_m\}$. Knowing some element of an API should generate an information gain formulated as

$$IG(S, API) = H(S) - H(S|API)$$

(2)

$$H(S|API) = \sum_{a_i} p(API = a_i) H(S|API = a_i)$$

(3)

A good API should order its elements according to the *law of diminishing returns*. This entails knowing that, for example, a package name should have a higher impact on your information gain (according to (2) and (3)) than stumbling across an internal method parameter of a deeply hidden class. Also, after seeing some function's name and establishing a sound understanding about its behavior, none of its parameters should oblige you to reevaluate your original conception. The next subsection will illuminate this issue.

There will always be emergent properties and empirically observable behaviors of a system not specified in an API. Such "holes" may be occluded via consumer-driven contracts (CDCs). These are augmentations of the API that are created by clients (they may be shaped as a set of test cases). Once those are attached to an API, then the cumulative information gain due to CDCs is defined as

$$IG(S, CDC) = IG(S, API \cup CDC)$$

(4)

All this semiformal reasoning entails that you must carefully name your abstractions. The names must not contradict the abstractions themselves. The next section gives a concrete example.

Fervent Flexibility Hurts Your API

Suppose that you have published your initial version of a package with all sorts of mathematical sequences. Among them is the function to produce the Fibonacci sequence with n elements, as shown in Listing 3-11.

Listing 3-11. Listing of fibonacci1.py to Produce the First n Fibonacci Numbers

```
def fibonacci(n):
    sequence = []
    current, next = 0, 1
    for _ in range(n):
        current, next = next, current + next
        sequence.append(current)
    return sequence
```

If you execute fibonacci(10), then you will get [1, 1, 2, 3, 5, 8, 13, 21, 34, 55]. The bold parts of Listing 3-11 constitute the contract for the abstraction called *Fibonacci sequence*. The name of the function tells us what we should expect as a return value given an input n. All seems consistent for now.

In the next release, you decide to make the function more flexible and allow clients to produce Fibonacci-like sequences by varying the starting condition. Listing 3-12 shows this new version of the fibonacci function.

Listing 3-12. Listing of fibonacci2.py to Produce the First n
Fibonacci-like Numbers

```
def fibonacci(n, f0=0, f1=1):
    sequence = []
    current, next = f0, f1
    for _ in range(n):
        current, next = next, current + next
        sequence.append(current)
    return sequence
```

If you execute fibonacci(10), then you will still get [1, 1, 2, 3,
5, 8, 13, 21, 34, 55], so compatibility is preserved. However, if you
run fibonacci(10, 5, 12), then you will receive [12, 17, 29, 46, 75,
121, 196, 317, 513, 830]. This sequence is correctly constructed by the
code but is pointless. It has nothing to do with the Fibonacci sequence.[3]
The function's name fibonacci is inconsistent with its behavior.
Furthermore, you must reassess your impression about the function after
seeing those strange parameters f0 and f1.

This problem is also abundant in popular frameworks. For example,
here is the signature of the function from NumPy to allegedly produce
a histogram: histogram(a, bins=10, range=None, **normed**=None,
weights=None, **density**=None). If they aren't set to True, then the output
is simply not a histogram (the normed parameter is deprecated, so only
density matters).

Listing 3-13 shows a much better design if there is still a need to
provide a generic sequencer function.

---

[3]For example, the Lucas sequence starts differently, although it has the same
recurrent formula as the Fibonacci sequence: L(n) = L(n - 1) + L(n - 2),
L(0) = 2, L(1) = 1.

Listing 3-13. Listing of sequencer.py to Produce Various Sequences

```
def simple_recurrent_sequence(n, first, second, combine_fun):
    sequence = []
    current, next = first, second
    for _ in range(n):
        current, next = next, combine_fun(current, next)
        sequence.append(current)
    return sequence

def fibonacci(n):
    return simple_recurrent_sequence(n, 0, 1, lambda x, y: x + y)
```

Now, any client may use fibonacci directly, or devise their own sequence with this new utility function. Even a more correct design would distribute these functions into separate modules. They are not at the same abstraction level. Grouping familiar abstractions in the same layer is the tenet of stratified or layered design. APIs should follow this strategy.

The Socio-* Pieces of Software Production

This section demonstrates from a different angle what happens when software development neglects the socio-technical and socio-economic aspects of software. The following excerpt nicely expresses the core problem:

> *Those trained and experienced in software development are often not trained in the kinds of socio-economic organizational skills needed for running large-scale software development. The socio-technical and socio-economic requirements for software production work remain understudied and elusive.*[4]

---

[4]Walt Scacchi, "Winning and Losing in Large-Scale Software Development: A Multi-Decade Perspective," *Computer* 51, no. 10 (October 2018): 58–65.

This conundrum is tightly associated with data science. Data science is a team sport, relies on software solutions, and executes in a broader sociological context (this is also the primary focus of systems engineering). Consequently, the following example equally applies both to systems and software engineering and to data science. The objective is to debunk the myth that being purely data-driven is both a necessary and sufficient condition for success.

There are two-primary socio-* parts of software: socio-economic and socio-technical. These have to be in balance. The socio-economic component ensures healthy business, while the socio-technical aspect keeps people happy. The crucial problem with data-driven approaches is that they are good for things that can be measured (preferably with tool support) but are not good for the many equally important characteristics that are hard to quantify.

Funny Elevator Case Study

Suppose you have a funny elevator that during the daytime ascends U(p) meters, while during the night descends D(own) meters. It must reach some total height H(eight) in meters. You need to implement the function num_days(h, u, d) to calculate the number of days for attaining the desired height. You can assume that $0 \leq D < U < H < 10^{30}$. For example, calling num_days(6, 5, 1) should yield 2. Listing 3-14 shows the initial version (see the elevator0.py file in the optimization folder). The running time is O(h / (u – d)) (for a good recap of Big-O notation, visit http://bigocheatsheet.com).

Listing 3-14. The First Unoptimized Variant of Our Function

```
def num_days(h, u, d):
    total_days = 1
    curr_height = 0
```

```
    while h - curr_height > u:
        curr_height += u - d
        total_days += 1
    return total_days
```

The preceding code literally simulates the climbing process with a conditional loop. Here are some timings of executions (notice the bold lines, which you shouldn't try to run):

```
>> %timeit num_days(6, 5, 1)
301 ns ± 7.52 ns per loop (mean ± std. dev. of 7 runs, 1000000
loops each)
>> %timeit num_days(1000000000, 4, 2)
>> %timeit num_days(998998998, 123461, 123460)
>> %timeit num_days(1000000000, 2314601, 2314600)
```

Clearly, the performance is unacceptable, and the software must be enhanced. Before any optimization work, you must unambiguously define what *fast* enough means. To aid testing, you must specify three values:

- **Minimum performance**: Below this threshold, the product is not acceptable. In our case, this value is 5 seconds for all test cases (i.e., the worst-case running time).

- **Target performance**: This is what you are aiming for. In our case, 1 second.

- **Maximum performance**: Above this threshold, there is economic motivation to spend resources for further improvement. In our case, 0.1 second.

Without this range, testers may reject your work just because execution time is negligibly below the target. Most of the time, the product is performing quite well even without exactly hitting the target. Of course, you must establish the baseline (our initial code) and track progress over time.

113

First Optimization Attempt

We can avoid many iterations by observing that the total number of days is at least $\left\lceil \dfrac{h}{u} \right\rceil$. We just need to account for the nightly plunges. Listing 3-15 shows the faster version of our code.

Listing 3-15. The Next Variant of Our Software

```
def num_days(h, u, d):
    total_days = 1
    height_left = h

    while u < height_left:
        days = height_left // u
        total_days += days
        height_left -= days * (u - d)
    return total_days
```

Following are the new execution times (the bold line is still out of reach of this program):

```
>> %timeit num_days(6, 5, 1)
321 ns ± 10.4 ns per loop (mean ± std. dev. of 7 runs, 1000000
loops each)
>> %timeit num_days(1000000000, 4, 2)
6.96 µs ± 438 ns per loop (mean ± std. dev. of 7 runs, 100000
loops each)
>> %timeit num_days(998998998, 123461, 123460)
235 ms ± 2.49 ms per loop (mean ± std. dev. of 7 runs, 1 loop each)
>> %timeit num_days(1000000000, 2314601, 2314600)
3.18 s ± 104 ms per loop (mean ± std. dev. of 7 runs, 1 loop each)
>> %timeit num_days(10 ** 20, 2314601, 2314600)
```

This is a remarkable improvement compared to the original version. After all, the first four test cases are within acceptable limits. Data (running times) should reassure us that we are on the right track. Observe that the code is a bit more complex, which is a natural consequence of performance optimization.

Note   We are now moving in the realm of socio-economic aspects, since we are trying to optimize a business-critical part of the code base. We measure improvements and let data drive us toward better variants. This is a typical route in mature software companies. You should avoid the temptation to make changes based on guessing about problems.

Second Optimization Attempt

It is possible to apply the *divide and conquer* paradigm, since we can combine the results from two subproblems of size $H/2$ to handle size H. This may drive us toward a logarithmic algorithm, provided that we find an efficient way to merge subcases. Listing 3-16 shows this approach in action; it applies many software engineering techniques: the divide and conquer paradigm, memorization, and hybrid approach.

Listing 3-16. The Very Fast Variant of Our Code

```
from functools import lru_cache

@lru_cache(maxsize=32)
def _partial_num_days(height_left, u, d):
    total_days = 1

    while u < height_left:
        days = height_left // u
```

```
        total_days += days
        height_left -= days * (u - d)
    return total_days

H_LIMIT = 1000000

def num_days(h, u, d):
    if h > H_LIMIT:
        days = 2 * (num_days(h // 2, u, d) - 1)
        height_left = h - days * (u - d)
        return days + _partial_num_days(height_left, u, d)
    else:
        return _partial_num_days(h, u, d)
```

Notice that the _partial_num_days function is essentially our previous num_days function. It covers low-dimensional cases. The threshold H_LIMIT controls when recursion should happen. This arrangement of selecting algorithms based on some condition is the hallmark of the hybrid approach. Below are the execution times:

```
>> %timeit num_days(6, 5, 1)
298 ns ± 14 ns per loop (mean ± std. dev. of 7 runs, 1000000
loops each)
>> %timeit num_days(1000000000, 4, 2)
5.6 µs ± 155 ns per loop (mean ± std. dev. of 7 runs, 100000
loops each)
>> %timeit num_days(998998998, 123461, 123460)
6.01 µs ± 56.9 ns per loop (mean ± std. dev. of 7 runs, 100000
loops each)
>> %timeit num_days(1000000000, 2314601, 2314600)
6.38 µs ± 998 ns per loop (mean ± std. dev. of 7 runs, 1 loop each)
>> %timeit num_days(10 ** 20, 2314601, 2314600)
29.9 µs ± 1.19 µs per loop (mean ± std. dev. of 7 runs, 1 loop each)
```

All running times are way below our imaginary 0.1 second. Should anyone question this achievement? Solely from the socio-economic viewpoint, this version is perfect. It is a business enabler, since all use cases are working within acceptable performance limits. What about the socio-technical aspect? This is exactly what many organizations miss due to blindly being data-driven. *This version is an abomination from the socio-technical perspective. It is simply awful!*

Teammate- and Business-Friendly Variant

List 3-17 shows the final version, which balances both socio-economic and socio-technical aspects of a solution. It is a straightforward, ultra-fast, and short program (the extra variables further clarify the partial expressions).

Listing 3-17. Overall Best Version of Our Code with O(1) Running Time

```
def num_days(h, u, d):
    import math

    height_left_until_last_day = h - u
    daily_progress = u - d
    return 1 + math.ceil(height_left_until_last_day / daily_
    progress)
```

Is it possible to land here after publishing the previous incomprehensible variant? Well, this is a tough question. You will need to convince management that something that enabled the business to roll now must be rewritten. Your only chance is to apply software economics and justify your intention by expressing how much the business will save through such maintainable software. It will be very hard to explain in monetary terms the beautiful code's positive impact on team morale and happiness. For this you must be a lucky person who is surrounded by unselfish management.

EXERCISE 3-1. DRAW THE CONE OF UNCERTAINTY

Look up the Cone of Uncertainty at `https://www.construx.com/books/the-cone-of-uncertainty` and try to reproduce it using the matplotlib framework (which is introduced in Chapter 6). The outcome of such a visualization would be a conceptual diagram. It doesn't contain actual values, but rather artificial ones just to showcase the phenomenon.

Hint: You may want to read the excellent tutorial for matplotlib at `https://realpython.com/python-matplotlib-guide`, which contains examples of conceptual graphs.

Summary

This chapter has tried to illuminate the importance of being context aware and knowing the sociological aspects of software production. Both are indispensable attributes in large-scale software development and data science. They cannot be treated separately once you recognize that computing is at the heart of scaling toward Big Data problems. Most topics discussed in this chapter are barely touched upon in classical courses, and thus practitioners typically learn them the hard way.

Python is the most popular language of data science thanks to its powerful frameworks, like those found in the SciPy ecosystem. Acquaintance with reuse-based software engineering is essential to properly apply those frameworks. Part of this effort revolves around topics touched upon in this chapter (APIs, readable code bases, adequate documentation, judicious use of abstractions, usage of tools, etc.).

As a data scientist, you will surely work under pressure with strict deadlines. This is the main difference between working as a professional in a software company and volunteering on open-source projects. This chapter's goal was to prepare you for working in large enterprises.

References

1. Liam Tung, "Developers, Despair: Half Your Time Is Wasted on Bad Code," `https://www.zdnet.com/article/developers-despair-half-your-time-is-wasted-on-bad-code`, Sept. 11, 2018.

2. Ervin Varga, *Unraveling Software Maintenance and Evolution: Thinking Outside the Box*, Springer, 2017.

3. Pierre Bourque and Richard E. (Dick) Fairley (Editors), *Guide to the Software Engineering Body of Knowledge (SWEBOK): Version 3.0*, IEEE Computer Society, 2014.

4. Ervin Varga, *Creating Maintainable APIs: A Practical, Case-Study Approach*, Apress, 2016.

5. Ervin Varga, "Do APIs Matter?," `https://www.apress.com/gp/blog/all-blog-posts/do-apis-matter/11524110`, Dec. 10, 2016.

CHAPTER 4

Documenting Your Work

Data science and scientific computing are human-centered, collaborative endeavors that gather teams of experts covering multiple domains. You will rarely perform any serious data analysis task alone. Therefore, efficient intra-team communication that ensures proper information exchange within a team is required. There is also a need to convey all details of an analysis to relevant external parties. Your team is part of a larger scientific community, so others must be able to easily validate and verify your team's findings. Reproducibility of an analysis is as important as the result itself. Achieving this requirement—to effectively deliver data, programs, and associated narrative as an interactive bundle—is not a trivial task. You cannot assume that everybody who wants to peek into your analysis is an experienced software engineer. On the other hand, all stakeholders aspire to make decisions based on available data. Fortunately, there is a powerful open-source solution for reconciling differences in individuals' skill sets. This chapter introduces the project *Jupyter* (see `https://jupyter.org`), the most popular ecosystem for documenting and sharing data science work.

The key to understanding the Jupyter project is to grasp the *Jupyter Notebook* architecture (see references [1–3]). A *notebook* in Jupyter is an interactive, executable narrative, which is buttressed by a web application running inside a browser. This web platform provides a convenient environment for doing all sorts of data science work, such as data

© Ervin Varga 2019
E. Varga, *Practical Data Science with Python 3*,
https://doi.org/10.1007/978-1-4842-4859-1_4

preprocessing, data cleaning, exploratory analysis, reporting, etc. It may also serve as a full-fledged development environment and supports a multitude of programming languages, including **Ju**lia, **Py**thon, **R**, and Scala.[1] As is emphasized on the Jupyter home page, the project "exists to develop open-source software, open-standards, and services for interactive computing across dozens of programming languages." The project provides the building blocks for crafting all kinds of custom interactive computing environments. Here are the three principal components of the Jupyter architecture:

> **Notebook document format**: A JSON document format for storing all types of content (code, images, videos, HTML, Markdown, LaTeX equations, etc.). The format has the extension .ipynb. See https:// github.com/jupyter/nbformat.

> **Messaging protocol**: A network messaging protocol based on JSON payloads for web clients (such as Jupyter Notebook) to talk to programming language kernels. A *kernel* is an engine for executing live code inside notebooks. The protocol specifies ZeroMQ or WebSockets as a transport mechanism. See https://github.com/jupyter/jupyter_client.

> **Notebook server**: A server that exposes WebSocket and HTTP Resource APIs for clients to remotely access a file system, terminal, and kernels. See https://github.com/jupyter/jupyter_server.

These elements are visualized in Figure 4-1. The web application may be the original Jupyter Notebook or some other compatible web user interface provider. The server process communicates with many different

---

[1]The name Jupyter originated as an amalgamation of the letters shown in bold from the supported languages.

kernels. Every notebook instance is associated with the matching kernel process. A notebook is comprised from many cells, each of which can be either text or live code. The code is executed by the proper language-dependent kernel. Code in a notebook may refer to previously defined code blocks. Usually, live code produces some read-only content, which becomes an integral part of the notebook. Such autogenerated content may be later browsed without starting any computation.

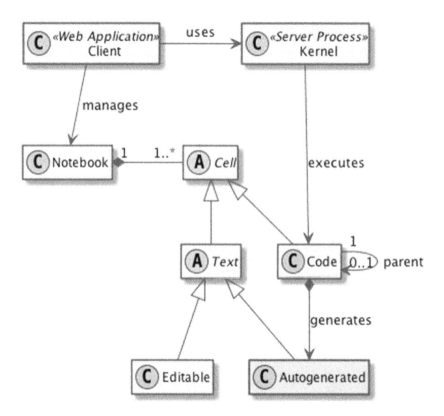

Figure 4-1. *The major components of the Jupyter architecture as well as the structural decomposition of a notebook. Each notebook is associated with its dedicated kernel at any given point in time.*

The various tools encompassed in the Jupyter ecosystem are as follows:

Jupyter Notebook: The first client/server stack of the architecture, which is still widely used at the time of writing.

JupyterLab: The next-generation web client application for working with notebooks.[2] We will implement it in this chapter. A notebook created with JupyterLab is fully compatible with Jupyter Notebook and vice versa.

JupyterHub: A cloud-based hosting solution for working with notebooks. It is especially important for enabling large organizations to scale their Jupyter deployments in a secure manner. Superb examples are UC Berkeley's "Foundations of Data Science" and UC San Diego's "Data Science MicroMasters" MOOC programs on the edX platform.

IPython: The Python kernel that enables users to use all extensions of the IPython console in their notebooks. This includes invoking shell commands prefixed with !. There are also magic commands for performing shell operations (such as %cd for changing the current working directory). You may want to explore the many IPython-related tutorials at https://github.com/ipython/ipython-in-depth.

---

[2]Another two popular web clients, with extensions to communicate with a notebook server, are *nteract* (see https://nteract.io/) and *Google Colaboratory* (see https://colab.research.google.com). There is also a tool called *Binder* (see https://mybinder.org) that can turn your GitHub repository with passive notebooks into a live interactive environment. It packages your notebooks into a Docker image amenable to being deployed into JupyterHub.

Jupyter widgets and notebook extensions: All sorts of web widgets to bolster interactivity of notebooks as well as extensions to boost your notebooks. We will demonstrate some of them in this chapter. For a good collection of extensions, visit `https://github.com/ipython-contrib/jupyter_contrib_nbextensions`.

nbconvert: Converts a notebook into another rich content format (e.g., Markdown, HTML, LaTeX, PDF, etc.). See `https://github.com/jupyter/nbconvert`.

nbviewer: A system for sharing notebooks. You can provide a URL that points to your notebook, and the tool will produce a static HTML web page with a stable link. You may later share this link with your peers. See `https://github.com/jupyter/nbviewer`.

JupyterLab in Action

You should start the *Anaconda Navigator* as appropriate to your operating system. Chapter 1 provides instructions for installing it on your operating system (you can also instantly try various Jupyter applications by visiting `https://jupyter.org/try`), and Figure 1-4 shows Anaconda Navigator's main screen with JupyterLab in the upper-left corner. Click JupyterLab's Launch button to start the tool. JupyterLab will present its main page (dashboard) inside your web browser. The screen is divided into three areas:

- The top menu bar includes commands to create, load, save, and close notebooks, create, delete, and alter cells, run cells, control the kernel, change views, read and update settings, and reach various help documents.

- The left pane has tabs for browsing the file systems, supervising the running notebooks, accessing all available commands, setting the properties of cells, and seeing what tabs are open in the right pane. The file browser's tab is selected by default and allows you to work with directories and files on the notebook server (this is the logical root). If you run everything locally (your server's URL will be something like `http://localhost:88xx/lab`), then this will be your home folder.

- The right pane has tabs for active components. The *Launcher* (present by default) contains buttons to fire up a new notebook, open an IPython console, open a Terminal window, and open a text editor. Newly opened components will be tiled in this area.

Experimenting with Code Execution

In the spirit of data science, let's first do some experiments with code execution. The goal is to get a sense of what happens when things go wrong, since being able to quickly debug issues increases productivity. Inside the file browser, click the toolbar button to create a new folder (the standard folder button with a plus sign on it). Right-click the newly created folder and rename it to `hanoi`. Double-click it to switch into that directory. Now, click the button in the Launcher in the right pane to open a notebook. Right-click inside the file browser on the new notebook file and rename it to `Solver1.ipubn`. If you have done everything properly, then you should see something similar to the screen shown in Figure 4-2.

Tip   If you have made an error, don't worry. You can always delete and move items by using the context menu and/or drag-and-drop actions. Furthermore, everything you do from JupyterLab is visible in your favorite file handler, so you can fix things from there, too. I advise you to always create a designated directory for your project. This avoids clutter and aids in organizing your artifacts. You will need to reference many other items (e.g., images, videos, data files, etc.) from your notebook. Keeping these together is the best way to proceed.

Enter into the code cell the erroneous function to solve the Hanoi Tower problem, as shown in Listing 4-1. Can you spot the problem without executing it? Execute the cell by clicking the Run button (large green arrow) on the toolbar (or simply press Shift+Enter). Observe that by running your cell in this fashion, JupyterLab automatically creates a new cell below it. This is useful when you continually add content to your document.

Listing 4-1. Hanoi Tower Solver with a Syntax Error

```
def solve_tower(num_disks, start, end, extra):
    if (num_disks > 0):
        solve_tower(num_disks - 1, start, extra, end)
        print('Move top disk from', start, 'to", end)
        solve_tower(num_disks - 1, extra, end, start)

solve_tower(3, 'a', 'c', 'b')
```

The output will be the following error message:

```
File "<ipython-input-2-7eeda1002555>", line 4
    print('Move top disk from', start, 'to", end)
                                                  ^
```

SyntaxError: EOL while scanning string literal

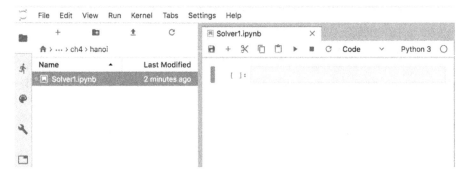

Figure 4-2. *The newly created notebook opened in the right pane*

The first cell is empty, and you may start typing in your code. Notice that its type is Code, which is shown in the drop-down box. Every cell is demarcated with a rectangle, and the currently active one has a thick bar on its left side. The field surrounded by square brackets is the placeholder for the cell's number. A cell receives a new identifier each time after being run. There are two types of numerated cells (see Figure 4-1): editable code cell (its content is preserved inside the In collection) and auto-generated cell (its content is preserved inside the Out collection). For example, you can refer to a cell's output by typing Out[X], where X is the cell's identifier. An immediate parent's output can be referenced via _, such as _X for a parent of X. Finally, the history is searchable and you may press the Up and Down keys to find the desired statement.

The error message in the output from Listing 4-1 is correct about encountering a syntax error. Nonetheless, the explanation is not that helpful, and is even misleading. Observe the bold characters in the error report. The caret symbol is Python's guess about the location of the error, while the real error is earlier. It is caused by an imbalanced string marker. You may use either " or ' to delineate a string in Python, but do not mix them for the same string.

To make this notebook a good reminder and worthy educational material, put the following Markdown content into the cell below the error output (don't forget to change the type of the cell to Markdown in the drop-down box):

This error message is an example that Python sometimes wrongly guesses the location of the error. **Different string markers should not be mixed for the same string**.

If you execute this cell (this time choose Run ➤ Run Selected Cells and Don't Advance to eschew creating a new cell), you will get nicely rendered HTML content. Your screen should look similar to Figure 4-3.

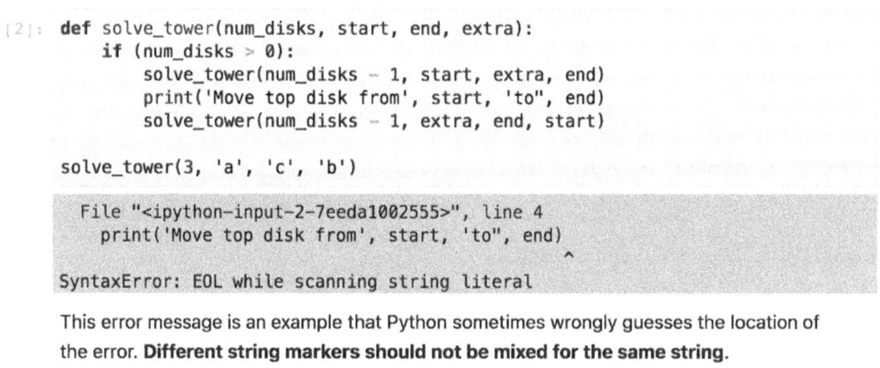

```
[2]: def solve_tower(num_disks, start, end, extra):
         if (num_disks > 0):
             solve_tower(num_disks - 1, start, extra, end)
             print('Move top disk from', start, 'to", end)
             solve_tower(num_disks - 1, extra, end, start)

     solve_tower(3, 'a', 'c', 'b')
       File "<ipython-input-2-7eeda1002555>", line 4
         print('Move top disk from', start, 'to", end)
                                                       ^
     SyntaxError: EOL while scanning string literal
```

This error message is an example that Python sometimes wrongly guesses the location of the error. **Different string markers should not be mixed for the same string.**

Figure 4-3. *Your first completed notebook showing an edge case of Python's error reporting. Notice that anyone can see all the details without running the code.*

Now save your notebook by clicking the disk icon button in the toolbar. Afterward, close the notebook's window and shut it down (choose the tab with a symbol of a runner in the left pane and select SHUTDOWN next to your notebook). Simply closing the UI page does not terminate the dedicated background process, so your notebook will keep running.

Repeat the earlier steps to create a new notebook and name it Solver2.ipubn. Enter the Hanoi Tower solver version shown in Listing 4-2 into a code cell and run it.

Listing 4-2. Hanoi Tower Solver with Infinite Recursion

```
def solve_tower(num_disks, start, end, extra):
    if (num_disks > 0):
        solve_tower(num_disks - 1, start, extra, end)
        print('Move top disk from', start, 'to', end)
        solve_tower(num_disks, extra, end, start)

solve_tower(3, 'a', 'c', 'b')
```

You will notice a very strange behavior. It will print an endless list of messages to move disks around, and finally report an error about reaching the maximum recursion depth. Instead of waiting for your code to blow up the stack, you may want to stop it. Such abrupt stoppage is the only option if your code enters an infinite loop. You can interrupt the code's execution by invoking Kernel ➤ Interrupt Kernel (note the many other kernel-related commands available in the Kernel menu). The effect of this action is visible in the following abridged output:

```
...
Move top disk from b to c
Move top disk from a to c
Move top disk from b to c
Move top disk from a to c
Move top disk from b to
-----------------------------------------------------------------
```

```
KeyboardInterrupt                           Traceback (most
recent call last)
<ipython-input-2-a4a17e313c43> in <module>()
      5              solve_tower(num_disks, extra, end, start)
      6
----> 7 solve_tower(3, 'a', 'c', 'b')

<ipython-input-2-a4a17e313c43> in solve_tower(num_disks, start,
end, extra)
      1 def solve_tower(num_disks, start, end, extra):
      2     if (num_disks > 0):
----> 3         solve_tower(num_disks - 1, start, extra, end)
      4         print('Move top disk from', start, 'to', end)
      5         solve_tower(num_disks, extra, end, start)
...
```

Now, enter the final variant of the program to solve the Tower of Hanoi puzzle, as shown in Listing 4-3. Of course, you should first repeat the previous steps and create a new notebook, Solver3.ipubn. This revision also contains embedded documentation explaining the purpose of the routine. Even though you may describe your code in narrative, it is always beneficial to document functions/routines separately. You can easily decide to move them from your notebook into a common place. Furthermore, input arguments are rarely described inside text cells, and it is easy to forget the details. All of this perfectly aligns with the following observation:

> *After all, the critical programming concerns of software engineering and artificial intelligence tend to coalesce as the systems under investigation become larger.*
>
> —Alan J. Perlis, Foreword to *Structure and Interpretation of Computer Programs, Second Edition* (MIT Press, 1996)

Listing 4-3. Correct Hanoi Tower Solver with Embedded
Documentation and Type Annotations

```python
def solve_tower(num_disks:int, start:str, end:str, extra:str)
-> None:
    """

    Solves the Tower of Hanoi puzzle.

    Args:
    num_disks: the number of disks to move.
    start: the name of the start pole.
    end: the name of the target pole.
    extra: the name of the temporary pole.

    Examples:
    >>> solve_tower(3, 'a', 'c', 'b')
    Move top disk from a to c
    Move top disk from a to b
    Move top disk from c to b
    Move top disk from a to c
    Move top disk from b to a
    Move top disk from b to c
    Move top disk from a to c
    >>> solve_tower(-1, 'a', 'c', 'b')
    """

    if (num_disks > 0):
        solve_tower(num_disks - 1, start, extra, end)
        print('Move top disk from', start, 'to', end)
        solve_tower(num_disks - 1, extra, end, start)
```

The code comment also incorporates doctest tests. These can be executed by running the following code cell (it is handy to put such a cell at the end of your notebook):

```
import doctest
doctest.testmod(verbose=True)
```

The output reflects the number and outcome of your tests:

```
Trying:
    solve_tower(3, 'a', 'c', 'b')
Expecting:
    Move top disk from a to c
    Move top disk from a to b
    Move top disk from c to b
    Move top disk from a to c
    Move top disk from b to a
    Move top disk from b to c
    Move top disk from a to c
ok
Trying:
    solve_tower(-1, 'a', 'c', 'b')
Expecting nothing
ok
1 items had no tests:
    __main__
1 items passed all tests:
    2 tests in __main__.solve_tower
2 tests in 2 items.
2 passed and 0 failed.
Test passed.
TestResults(failed=0, attempted=2)
```

Finally, such documentation can be easily retrieved by executing `solve_tower?` (if you include two question marks, then you can dump the source code, too).

Managing the Kernel

The kernel is an engine that runs your code and sends back the results of execution. In Figure 4-2 (in the previous section), you can see the kernel's status; look at the circle in the upper-right corner, next to the type of the kernel (in our case, Python 3). A white circle means that the kernel is idle and is ready to execute code. A filled-in circle denotes a busy kernel (this same visual clue also designates a dead kernel). If you notice that your program is halted, or the interaction with your notebook becomes tedious (slow and unresponsive behavior), then you may want to consider the following commands (all of them are grouped under the Kernel menu, and each menu item with an ellipsis will open a dialog box for you to confirm your intention):

> **Interrupt Kernel**: Interrupts your current kernel. This is useful when all system components are healthy except your currently running code (we have already seen this command in action).

> **Restart Kernel...**: Restarts the engine itself. You should try this if you notice sluggish performance of your notebook.

> **Restart Kernel and Clear...**: Restarts the server and clears all autogenerated output.

> **Restart Kernel and Run All...**: Restarts the server and runs all code cells from the beginning of your notebook.

Shutdown Kernel: Shuts down your current kernel.

Shutdown All Kernels...: Shuts down all active kernels. This applies when your notebook contains code written in different supported programming languages.

Change Kernel...: Changes your current kernel. You can choose a new kernel from the drop-down list. One of them is the dummy kernel called No Kernel (with this, all attempts to execute a code cell will simply wipe out its previously generated output). You can also choose a kernel from your previous session.

Note   Whenever you restart the kernel, you must execute your code cells again in proper order. Forgetting this is the most probable cause for an error like `NameError: name <XXX> is not defined`.

Connecting to a Notebook's Kernel

If you would like to experiment with various variants of your code without disturbing the main flow of your notebook, you can attach another front end (like a terminal console or Qt Console application) to the currently running kernel. To find out the necessary connection information, run inside a code cell the `%connect_info` magic command.[3] The output will also give you some hints about how to make a connection. The nice thing about this is that you will have access to all artifacts from your notebook.

---

[3]The fastest way is to just run `%qtconsole` inside a code cell, and this will summon a Qt Console tool attached to your kernel. The command will automatically pick up the required connection information.

135

Caution   Make sure to always treat your notebook as a source of truth. You can easily introduce a new variable via your console, and it will appear as defined in your notebook, too. Don't forget to put that definition back where it belongs, if you deem it to be useful.

Descending Ball Project

We will now develop a small but complete project to showcase other powerful features of JupyterLab (many of them are delivered by IPython, as this is our kernel). The idea is to make the example straightforward so that you can focus only on JupyterLab. Start by creating a new folder for this project (name it ball_descend). Inside it create a new notebook, Simulation.ipubn. Revisit the "Experimenting with Code Execution" section for instructions on how to accomplish these steps.

Problem Specification

It is useful to start your notebook with a clear title and a description of the problem. These details are very important to highlight from the very beginning the essence of your work. Don't let others waste time trying to figure out whether your work is valuable to them or not. Select the Markdown cell type from the drop-down box, and enter the following text:

```
# Simulation of a Bal's Descent in a Terrain
```

```
This project simulates where a ball will land in a terrain.
```

```
## Input
The terrain's configuration is given as a matrix of integers
representing elevation at each spot. For simplicity, assume
that the terrain is surrounded by a rectangular wall that
```

prevents the ball from escaping. The inner dimensions of the terrain are NxM, where N and M are integers between 3 and 1000.

The ball's initial position is given as a pair of integers (a, b).

Output
The result is a list of coordinates denoting the ball's path in a terrain. The first element of the list is the starting position, and the last one is the ending position. It could happen that they are the same, if the terrain is flat or the ball is dropped at the lowest point (the local minima).

Rules
The ball moves according to the next two simple rules:
- The ball rolls from the current position into the lowest neighboring one.
- If the ball is surrounded by higher points, then it stops.

You should utilize the rich formatting options provided by the Markdown format. Here, we define headers to create some structure in our document. We will talk more about structuring in the next section. Once you execute this cell, it will be rendered as HTML.

It turns out that the title contains a small typo (the issue is more apparent in the rendered HTML); it says Bal's instead of Ball's. Double-click the text cell, correct the problem, and rerun the cell.

You also could have entered the preceding text using multiple consecutive cells. It is possible to split, merge, and move around cells by using the commands from the Edit menu or by using drag-and-drop techniques. The result of running a text cell is its formatted output. Text cells may be run in any order. This is not the case with code cells. They usually have dependencies on each other. Forgetting to run a cell on which your code depends may cause all sorts of errors. Obviously, reducing coupling between code cells is vital. A graph of dependencies between code cells may reveal a lot about complexity. This is one reason why you should minimize dependencies on global variables, as these intertwine your cells.

Model Definition

The problem description in the previous section adequately suggests the data model. One intuitive choice is to represent the terrain as a matrix of integers. *NumPy* already has such a data structure. Of course, for this miniature example, we could have used only standard Python stuff. Nonetheless, I want to give some hints about setup code. In the next code cell enter the content of Listing 4-4 and execute the cell (this will also create a new code cell below it).

Listing 4-4. Global Imports for the Notebook with a Comment for the User to Just Run It

```
# Usual bootstrapping code; just run this cell.
import numpy as np
```

Your notebook should be carefully organized with well-defined sections. Usually, bootstrapping code (such as shown in Listing 4-4) should be kept inside a single code cell at the very beginning of your notebook. Such code is not inherently related to your work, and thus should not be spread out all over your notebook. Moreover, most other code cells depend on this cell to be executed first. If you alter this section, then you should rerun all dependent cells. A handy way to accomplish this is to invoke Run ➤ Run Selected Cell and All Below.

Type into the code cell the content of Listing 4-5 and run it.

Listing 4-5. Definition of Our Data Model

```
terrain = np.matrix([
    [-2, 3, 2, 1],
    [-2, 4, 3, 0],
    [-3, 3, 1, -3],
    [-4, 2, -1, 1],
    [-5, -7, 3, 0]
])
terrain
```

A cell may hold multiple lines and expressions. Such a composite cell executes by sequentially running the contained expressions (in the order in which they appear). Here, we have an assignment and a value expression. The latter is useful to see the effect of the assignment (assignments are silently executed). Remember that a cell's output value is always the value of its last expression (an assignment has no value). If you want to dump multiple messages, you can use Python's print statement (these messages will not count as output values) with or without a last expression. Typically, an output value will be nicely rendered into HTML, which isn't the case with printed output. This will be evident when we output as a value a Pandas data frame in Chapter 5.

It is also possible to prevent outputting the last expression's value by ending it with a semicolon. This can be handy in situations where you just want to see the effects of calling some function (most often related to visualization).

When your cursor is inside a multiline cell, you can use the Up and Down arrows on your keyboard to move among those lines. To use your keys to move between cells, you must first escape the block by pressing the Esc key. It is also possible to select a whole cell by clicking inside an area between the left edge of a cell and the matching thick vertical bar. Clicking the vertical bar will shrink or expand the cell (a squashed cell is represented with three dots).

139

> **Tip**   It is possible to control the rendering mechanism for output values by toggling *pretty printing* on and off. This can be achieved by running the `%pprint` magic command. For a list of available magic commands, execute `%lsmagic`.

> **Note**   Never put inside the same cell a slow expression that always results in the same value (like reading from a file) and an idempotent parameterized expression (like showing the first couple of elements of a data frame). The last expression cannot be executed independently (i.e., without continuously reloading the input file).

The `matrix` function is just one of many from the NumPy package. JupyterLab's context-sensitive typing facility can help you a lot. Just press the Tab key after `np.`, and you will get a list of available functions. Further typing (for example, pressing the M key) will narrow down the list of choices (for example, to names starting with `m`). Moreover, issuing `np.matrix?` in a code cell provides you with help information about this function (you must execute the cell to see the message). Executing `np?` gives you help about the whole framework.

Path Finder's Implementation

We are now ready to tackle the essential piece of our project, the function to calculate the ball's path. The input arguments will be the terrain's configuration and the starting position of the ball. The output will be the list of coordinates that the ball would follow in the terrain. Figure 4-4 depicts the top-down decomposition of the initial problem into subproblems. Each subproblem is implemented as a separate function. We will start with the `wall` function (see Listing 4-6).

Listing 4-6. Definition of the *wall* Function to Detect Borders

```
def wall(terrain:np.matrix, position:Tuple[int,int]) -> bool:
    """

    Checks whether the provided position is hitting the wall.

    Args:
    terrain: the terrain's configuration comprised from integer
    elevation levels.
    position: the pair of integers representing the ball's
    potential position.

    Output:
    True if the position is hitting the wall, or False
    otherwise.

    Examples:
    >>> wall(np.matrix([[-2, 3, 2, 1]]), (0, 1))
    False
    >>> wall(np.matrix([[-2, 3, 2, 1]]), (-1, 0))
    True
    """

    x, y = position
    length, width = terrain.shape
    return (x < 0) or (y < 0) or (x >= length) or (y >= width)
```

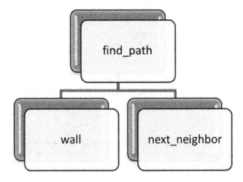

Figure 4-4. *Top-down decomposition of our problem; we will implement the functions via the bottom-up method*

The logic is really simple. The `wall` function's signature uses Python's optional type annotations. The `Tuple` must be imported by adding the next line into our bootstrapping cell (it must be rerun after the modification):

```
from typing import Tuple
```

I suggest that you be pragmatic with these annotations. For example, it is enough to state that the `terrain` is `np.matrix`, instead of embarking on custom type definitions to describe it in more detail. The next two functions in Listing 4-7 should be added inside the same cell with the `wall` function. It makes sense to keep them together because they are interrelated. Moreover, they can be hidden in an *all or nothing* fashion by clicking the cell's vertical bar.

Listing 4-7. Implementation of the Other Two Functions As Shown in Figure 4-4

```
def next_neighbor(terrain:np.matrix, position:Tuple[int,int])
-> Tuple[int,int]:
    """

    Returns the position of the lowest neighbor.
```

```
Args:
terrain: the terrain's configuration comprised from integer
elevation levels.
position: the pair of integers representing the ball's
current position.

Output:
The position (pair of coordinates) of the lowest neighbor.

Example:
>>> next_neighbor(np.matrix([[-2, 3, 2, 1]]), (0, 1))
(0, 0)
"""

x, y = position
allowed_neighbors = []
for delta_x in range(-1, 2):
    for delta_y in range(-1, 2):
        new_position = (x + delta_x, y + delta_y)
        if (not wall(terrain, new_position)):
            allowed_neighbors.append((terrain.item(new_
                position), new_position))
    return min(allowed_neighbors)[1]
def find_path(terrain:np.matrix, position:Tuple[int,int]) ->
List[Tuple[int,int]]:
    """

    Find the path that the ball would follow while descending
    in the terrain.

    Args:
    terrain: the terrain's configuration comprised from integer
    elevation levels.
    position: the pair of integers representing the ball's
    current position.
```

```
Output:
The list of coordinates of the path.

Example:
>>> find_path(np.matrix([[-2, 3, 2, 1]]), (0, 1))
[(0, 1), (0, 0)]
"""

next_position = next_neighbor(terrain, position)
if (position == next_position):
    return [position]
else:
    return [position] + find_path(terrain, next_position)
```

The find_path function is a very simple recursive function. The exit condition is the guaranteed local minima (unless you model a terrain from Escher's world), since we monotonically descend toward the lowest neighbor.

We must augment our list of imports in the setup section to include List, too. We must also add the following code cell to test our functions:

```
# Just run this cell to invoke tests embedded inside function
descriptors.
import doctest
doctest.testmod(verbose=True)
```

After executing all cells, we should receive a test result with no errors. Notice that all functions are self-contained and independent from the environment. This is very important from the viewpoint of maintenance and evolution. Interestingly, all functions would execute perfectly even if you were to remove the terrain argument from their signature. Nonetheless, dependence on global variables is an equally bad practice in notebooks as it is anywhere else. It is easy to introduce unwanted side-effects and pesky bugs. Nobody has time, nor incentive, to debug your document to validate your results!

Interaction with the Simulator

It is time to wrap up the project by offering to users a comfortable way to interact with our simulator. A classical approach would be to just create a code cell of the following form (each time a user would need to change the code and rerun the cell):

```
start_position = (1, 1)
find_path(terrain, start_position)
```

There is a much better way by exploiting *Jupyter Widgets*. First, you need to augment the bootstrapping cell with the following import:

```
from ipywidgets import interact, widgets
```

JupyterLab doesn't allow you to directly embed JavaScript-generated content into your document. You must install IPyWidgets as a JupyterLab extension. Save your notebook, shut down the kernel, and open a Terminal window. From a command line, execute the following command:

```
jupyter labextension install @jupyter-widgets/jupyterlab-manager
```

You will get a warning if you don't have NodeJS installed. You can easily install it by summoning

```
conda install nodejs
```

As explained in Chapter 1, these additions are best handled by first creating a custom environment for your project. Now start JupyterLab and load your notebook. Enter the following content into a new code cell:

```
interact(lambda start_x, start_y: find_path(terrain, (start_x,
start_y)),
        start_x = widgets.IntSlider(value=1, max=terrain.
        shape[0]-1, description='Start X'),
        start_y = widgets.IntSlider(value=1, max=terrain.
        shape[1]-1, description='Start Y'));
```

After you execute this cell, you will be presented with two named sliders to set the ball's initial position (X represents the row and Y the column). Each time you move the slider, the system will output a new path. There is no chance to provide an invalid starting position, as the sliders are configured to match the terrain's shape. The notebook included in this book's source code bundle also contains some narrative for presenting the result inside a separate section.

Test Automation

In this modern DevOps era, we cannot afford to perform tasks manually all the time. It is easy to open a notebook and select the menu item to run all cells. However, doing this repeatedly and frequently isn't feasible. We must be able to automate the whole process. Part of the build automation is to test whether all cells are appropriate in a notebook (in this manner, we can indirectly run doctest tests, too). Such a statement would be part of a build script. Open a Terminal window; this time do it from JupyterLab by clicking the corresponding button in the Launcher (if the Launcher tab isn't visible, choose File ➤ New Launcher). Ensure that you are inside the source code folder of this chapter. Execute the following statement[4]:

```
python -m pytest --nbval-lax ball_descend/Simulation.ipynb
```

You should see the following output:

```
=============== test session starts ==========================
platform darwin -- Python 3.6.5, pytest-3.6.0, py-1.5.3,
pluggy-0.6.0
rootdir: /Users/evarga/Projects/pdsp_book/src/ch4, inifile:
plugins: remotedata-0.3.0, openfiles-0.3.0, doctestplus-0.1.3,
arraydiff-0.2, nbval-0.9.1
```

---

[4]Visit https://github.com/computationalmodelling/nbval for instructions on how to install the *Py.test* plug-in for validating Jupyter notebooks.

```
collected 6 items

ball_descend/Simulation.ipynb
......                                                         [100%]

============ 6 passed in 2.15 seconds ========================
```

On the other hand, try to execute the following statement:

```
python -m pytest --nbval-lax hanoi/Solver1.ipynb
```

The tool will report an error, which is expected, as the notebook contains a code cell with a syntax error:

```
=============== test session starts ========================
platform darwin -- Python 3.6.5, pytest-3.6.0, py-1.5.3,
pluggy-0.6.0
rootdir: /Users/evarga/Projects/pdsp_book/src/ch4, inifile:
plugins: remotedata-0.3.0, openfiles-0.3.0, doctestplus-0.1.3,
arraydiff-0.2, nbval-0.9.1
collected 1 item

hanoi/Solver1.ipynb F                                        [100%]


==================== FAILURES ===============================
_____ hanoi/Solver1.ipynb::Cell 0 _____
Notebook cell execution failed
Cell 0: Cell execution caused an exception

Input:
def solve_tower(num_disks, start, end, extra):
    if (num_disks > 0):
        solve_tower(num_disks - 1, start, extra, end)
        print('Move top disk from', start, 'to", end)
        solve_tower(num_disks - 1, extra, end, start)
```

```
solve_tower(3, 'a', 'c', 'b')
```

Traceback:

```
  File "<ipython-input-1-7eeda1002555>", line 4
    print('Move top disk from', start, 'to", end)
                                                  ^
SyntaxError: EOL while scanning string literal
```

```
============ 1 failed in 2.16 seconds =========================
```

Refactoring the Simulator's Notebook

You should always seek to improve the clarity and structure of your artifacts. A notebook isn't an exception. The current one contains lots of Python code, which may distract the user from the main points of the work. We will make the following improvements:

1. Move out the wall, next_neighbor, and find_path functions into a separate package called pathfinder.

2. Move the doctest call into our new package.

3. Import the new package into our notebook (we need to access the find_path function).

4. Add more explanation about what we are doing, together with a nicely formatted formula.

Create a new folder named pathfinder in the current project folder. Create a new file in pathfinder and name it pathfinder.py. Copy the wall, next_neighbor, and find_path functions into this file. Remove the matching code cell from the notebook. Copy the imports of numpy and typing (located at the beginning of the notebook). Delete the typing import from the notebook. Create a file named __init.py__ in pathfinder (see reference [4]) and insert the following line:

```
from pathfinder.pathfinder import find_path
```

Move over the content of the code cell from the notebook that invokes doctest and put it in the following if statement at the end of pathfinder.py:

```
if __name__ == "__main__":
```

At the beginning of the notebook, insert the following import statement:

```
from pathfinder import find_path
```

At this point, your notebook should function the same as before. Notice its tidiness. Finally, add the following text to be the second and third sentences in your notebook:

It simulates the influence of Newton's law of universal gravitation on the movement of a ball, given by the formula $F=g\frac{m_1m_2}{r^2}$. Here, F is the resulting gravitational pull between the matching objects, m_1 and m_2 are their masses, r is the distance between the centers of their masses, and g is the gravitational constant.

The bold parts are LaTex expressions (they must be demarcated by $). After you execute the text cell, the formula will be nicely rendered, as shown in Figure 4-5. More complex LaTex content can be put inside the Raw cell type and rendered via the nbconvert command-line utility.

This project simulates where a ball will land in a terrain. It simulates the influence of Newton's law of universal gravitation on the movement of a ball, given by the formula $F = g\frac{m_1m_2}{r^2}$. Here, F is the resulting gravitational pull between the matching objects, m_1 and m_2 are their masses, r is the distance between the centers of their masses, and g is the gravitational constant.

Figure 4-5. *The LaTex content inside ordinary text is beautifully rendered into HTML*

Document Structure

In our previous project, we have already tackled the topic of content structuring, although in a really lightweight fashion. We will devote more attention to it here. Whatever technology you plan to use for your documentation task, you need to have a firm idea of how to structure your document. A structure brings order and consistency to your report. There are no hard rules about this structure, but there are many heuristics (a.k.a. best practices). The document should be divided into well-defined sections arranged into some logical order. Each section should have a clear scope and volume; for example, it makes no sense to devote more space to the introduction than to the key findings in your analysis. Remember that a notebook is also a kind of document and it must be properly laid out. Sure, the data science process already advises how and in what order to perform the major steps (see Chapter 1), and this aspect should be reflected in your notebook. Nonetheless, there are other structuring rules that should be superimposed on top of the data science life cycle model.

One plausible high-level document template may look as follows (I assume that a sound title/subtitle is mandatory in all scenarios):

> **Abstract**: This section should be a brief summary of your work. It must illuminate what you have done, in what way, and enumerate key results.

> **Motivation**: This section should explain why your work is important and how it may impact the target audience.

> **Dataset**: This section should describe the dataset and its source(s). You should give unambiguous instructions that explain how to retrieve the dataset for reproducibility purposes.

Data Science Life Cycle Phases: The next couple of sections should follow the data science life cycle model (date preprocessing, research questions, methods, data analysis, reporting, etc.) and succinctly explain each phase. These details are frequently present in data analysis notebooks.

Drawbacks: This section should honestly mention all limitations of your methodology. Not knowing about constraints is very dangerous in decision making.

Conclusion: This section should elaborate about major achievements.

Future Work: This section should give some hints about what you are planning to do in the future. Other scientists are dealing with similar issues, and this opens up an opportunity for collaboration.

References: This section should list all pertinent references that you have used during your research. Don't bloat this section as an attempt to make your work more "convincing."

The users of your work may be segregated into three major categories: the general public, decision makers, and technically savvy users. The general public is only interested in what you are trying to solve. Users in this category likely will read only the title and abstract. The decision makers are business people and are likely to read the major findings as well as the drawbacks and conclusion. They mostly seek well-formulated actionable insights. The technical people (including CTOs, data scientists, etc.) would also like to reproduce your findings and extend your research.

Therefore, they will look into all aspects of your report, including implementation details. If you fail to recognize and/or address the needs of these various classes of users, then you will reduce the potential to spread your results as broadly as possible.

Wikipedia Edits Project

As an illustration of the template outlined in the previous section, I will fill out some of the sections based upon my analysis of Wikipedia edits. The complete Jupyter notebook is publicly available at Kaggle (see `https://www.kaggle.com/evarga/analysis-of-wikipedia-edits`). It does contain details about major data science life cycle phases. The goal of this project is to spark discussion, as there are many points open for debate. The following sections from the template should be enough for you to grasp the essence of this analysis (without even opening the previously mentioned notebook). Don't worry if the Kaggle notebook seems complicated at this moment.

Abstract

This study uses the Wikipedia Edits dataset from Kaggle. It tries to inform the user whether Wikipedia's content is stable and accountable. The report also identifies which topics are most frequently edited, based on words in edited titles. The work relies on various visualizations (like scatter plot, stacked bar graphs, and word cloud) to drive conclusions. It also leverages NLTK to process the titles. We may conclude that Wikipedia is good enough for informal usage with proper accountability, and themes like movies, sports, and music are most frequently updated.

Motivation

Wikipedia often is the first web site that people visit when they are looking for information. Obviously, high quality (accuracy, reliability, timeliness, etc.) of its content is imperative for a broad community. This work tries to peek under the hood of Wikipedia by analyzing the edits made by users. Wikipedia can be edited by anyone (including bots), and this may raise concerns about its trustworthiness. Therefore, by getting more insight about the changes, we can judge whether Wikipedia can be treated as a reliable source of information. As a side note, scientific papers cannot rely on it. There are also some book publishers who forbid referencing Wikipedia. All in all, this report tries to shed light on whether Wikipedia is good enough for informal usage.

Drawbacks

- The data reflects an activity of users over a very short period of time. Such a small dataset cannot provide a complete story. Moreover, due to time zone differences, it cannot represent all parts of the world.

- There is no description on Kaggle about the data acquisition process for the downloaded dataset. Consequently, the recorded facts should be taken with a pinch of salt. The edit's size field is especially troublesome.

- The data has inherent limitations, too. I had no access to the user profiles, so I assumed all users are equally qualified to make edits. If I would have had this access, then I could have weighted the impact of their modifications.

Conclusion

- Wikipedia is good enough for informal usage. The changes are mostly about fixing smaller issues. Larger changes in size are related mostly to addition of new content and are performed by humans. These updates are treated as major.

- Larger edits are done by registered users, while smaller fixes are performed also by anonymous persons.

- Specialized content (scientific, technical/technology related, etc.) doesn't change as frequently as topics about movies, sport, music, etc.

EXERCISE 4-1. EXTERNAL LOAD OF DATA

Manually entering huge amounts of data doesn't scale. In our case study, the terrain's configuration fits into a 5×4 matrix. Using this approach to insert elevations for a 200×300 terrain would be impossible. A better tactic is to store data externally and load it on demand. Modify the terrain's initialization code cell to read data from a text file. Luckily, you don't need to wrestle with this improvement too much. NumPy's `Matrix` class already supports data as a string. Here is an excerpt from its documentation: "If `data` is a string, it is interpreted as a matrix with commas or spaces separating columns, and semicolons separating rows."

You would want to first produce a text file with the same content as we have used here. In this way, you can easily test whether everything else works as expected. You should upload the configuration file into the same folder where your notebook is situated (to be able to use only the file name as a reference). To upload stuff, click the `Upload Files` toolbar button in JupyterLab's file browser. Refer to Chapter 2 for guidance on how to open/read a text file in Python.

EXERCISE 4-2. FIXING SPECIFICATION AMBIGUITIES

Thanks to the accessibility of your JupyterLab notebook and the repeatability of your data analysis, one astute data scientist has noticed a flaw in your solution. He reported this problem with a complete executable example (he shared with you a modified notebook file). Hence, you can exactly figure out what he would like you to fix. For a starting position (0, 1) it is not clear in advance whether the ball should land in (0, 0) or (1, 0), since they are both at a locally lowest altitude (in this case -2). It is also not clear where the ball should go if it happens to land on a plateau (an area of the terrain at the same elevation). In the current solution, it will stop on one of the spots, depending on the search order of neighbors. Surely, this doesn't quite satisfy the rule of stopping at the local minima.

The questions are thus: Should you consider inertia? How do you document what point will be the final point? Think about these questions and expand the text and/or modify the path-finding algorithm.

EXERCISE 4-3. EXTENDING THE PROJECT

Another data scientist has requested an extension of the problem. She would like to ascribe elasticity to the ball. If it drops more than X units, then it could bounce up Y units. Change the path-finding algorithm to take this flexibility into account. Assume that the ball will still select the lowest neighbor, although the set of candidates will increase. Will the recursion in find_path always terminate? What conditions dictate such guaranteed termination? Test your solution thoroughly.

EXERCISE 4-4. NOTEBOOK PRESENTATION

In the "Document Structure" section, you can find the proposed document template. Creating a separate artifact, external to your main notebook, isn't a good choice, since it will eventually drift away from it (like passive design documents in software development).

A JupyterLab notebook can have a dual purpose: as presentation material and as an executable narrative. Extend the descending ball project's notebook with parts from the document template (add Abstract and Conclusion sections, for a start). Set the slide type of these textual cells to Slide. Mark other cells as Skip. Open a Terminal window and type:

```
jupyter nbconvert <YourSlide>.ipynb --to slides --post serve
```

A new browser window will open, presenting one after another cells marked as Slide. Look up the meaning of other slide types: Sub-slide, Fragment, and Notes. Consult JupyterLab's documentation for more information about presentation mode. For a really professional presentation, you should use *Reveal.js* (see https://revealjs.com).

Summary

This chapter covered some of the benefits of packaging documentation as a self-contained, executable, and shareable asset:

- Freeform text is bundled together with executable code; this eliminates the need to maintain separate documents, which usually get out of sync with code.

- The output of code execution may be saved in the document and become an integral part of it.

- The setup of an executable environment (to bring in dependencies for running your code) may be done inside the document. This solves many deployment problems and eliminates a steep learning curve for those who would like to see your findings in action.

You have witnessed the power behind computational notebooks and how the Jupyter toolset accomplishes most requirements regarding documentation. By supporting disparate programming languages, JupyterLab fosters polyglot programming, which is important in the realm of data science. In the same way as multiple data sources are invaluable, many differently optimized development/executable environments are indispensable in crafting good solutions.

JupyterLab has many more useful features not demonstrated here. For example, it has a web-first code editor that eliminates the need for a full-blown integrated development environment (such as Spyder) for smaller edits. You can edit Python code online far away from your machine. JupyterLab also allows you to handle data without writing Python code. If you open a Leaflet GeoJSON file (see `https://leafletjs.com`), then it will be immediately visualized and ready for interaction. A classical approach entails running a Python code cell.

All in all, this chapter has provided the foundation upon which further chapters will build. We will continually showcase new elements of JupyterLab, as this will be our default executable environment.

References

1. Brian Granger, "Project Jupyter: From Computational Notebooks to Large Scale Data Science with Sensitive Data," ACM Learning Seminar, September 2018.

2. Matt Cone, *The Markdown Guide*, `https://www.`
    `markdownguide.org.`

3. Jake VanderPlas, *Python Data Science Handbook:
    Essential Tools for Working with Data*, O'Reilly
    Media, Inc., 2016.

4. Mike Grouchy, "Be Pythonic: __init__.py," `https://`
    `mikegrouchy.com/blog/be-pythonic-__init__py`,
    May 17, 2012.

CHAPTER 5

Data Processing

Data analysis is the central phase of a data science process. It is similar to the construction phase in software development, where actual code is produced. The focus is on being able to handle large volumes of data to synthesize an actionable insight and knowledge. Data processing is the major phase where math and software engineering skills interplay to cope with all sorts of scalability issues (size, velocity, complexity, etc.). It isn't enough to simply pile up various technologies in the hope that all will auto-magically align and deliver the intended outcome. Knowing the basic paradigms and mechanisms is indispensable. This is the main topic of this chapter: to introduce and exemplify pertinent concepts related to scalable data processing. Once you properly understand these concepts, then you will be in a much better position to comprehend why a particular choice of technologies would be the best way to go.

Augmented Descending Ball Project

We will expand on the example from the previous chapter. As a reminder, here is the original problem specification in Markdown format (the bold section will be altered):

```
# Simulation of a Ball's Descent in a Terrain
```

```
This project simulates where a ball will land in a terrain.
```

© Ervin Varga 2019
E. Varga, *Practical Data Science with Python 3*,
https://doi.org/10.1007/978-1-4842-4859-1_5

Input

The terrain's configuration is given as a matrix of integers representing elevation at each spot. For simplicity, assume that the terrain is surrounded by a rectangular wall that prevents the ball from escaping. The inner dimensions of the terrain are NxM, where N and M are integers between 3 and 1000.

The ball's initial position is given as a pair of integers (a, b).

Output

The result is a list of coordinates denoting the ball's path in a terrain. The first element of the list is the starting position, and the last one is the ending position. It could happen that they are the same, if the terrain is flat or the ball is dropped at the lowest point (the local minima).

Rules

The ball moves according to the next two simple rules:
- The ball rolls from the current position into the lowest neighboring one.
- If the ball is surrounded by higher points, then it stops.

Version 1.1

Data science is all about combining different data sources in a meaningful way (dictated by business drivers). To make our product more useful for a broader community, we will extend the initial system to accept as input some real terrain's elevation data encoded in an image; open the Terrain_ Simulation_v1.1.ipubn notebook in the augmented_ball_descend source folder of this chapter. The rest of the specification should remain valid. Figure 5-1 presents one publicly available satellite image dataset from the WIFIRE Lab (visit https://wifire.ucsd.edu), which is also used in UC San Diego's MOOC Python for Data Science on edX. The WIFIRE system gets these images from an external source, NASA's Global Imagery Browse Services (GIBS) (see https://earthdata.nasa.gov, where GIBS is accessible from the DATA tab).

Figure 5-1. *The satellite image of a coastline with encoded terrain data; each pixel's red-green-blue components represent altitude, slope, and aspect, respectively. Higher color component values denote higher altitude, slope, and aspect.*

The following is the new description of what we expect as input (notice how coloring is done in Markdown):

The terrain's configuration is given as a matrix of integers representing elevation at each spot. This matrix is computed from a satellite image of a terrain that encodes altitude

161

inside a pixel's ***RED*** color component. For simplicity, assume that the terrain is surrounded by a wall, that prevents the ball to escape. The inner dimensions of the terrain are NxM, where N and M are integers.

The ball's initial position is given as a pair of integers (a, b).

The following snippet of code reads the image and dumps some statistics (the following couple of imports have been spread out over the notebook to highlight the changes):

```
import imageio
```

```
terrain = imageio.imread('terrain_data/coastline.jpg')
print("The terrain's dimensions are %s with minimum %i, maximum %i, and mean %.2f for points."%\
    (terrain.shape, terrain.min(), terrain.max(), terrain.mean()))
```

Running this cell produces the following output:

```
The terrain's dimensions are (3725, 4797, 3) with minimum 0,
maximum 255, and mean 75.83 ➡
for points.
```

So, the size of the input is increased from 5×4 to 3725×4797×3. The following lines display the image as depicted in Figure 5-1:

```
%matplotlib inline
import matplotlib.pyplot as plt

plt.figure(figsize=(13, 15))
plt.imshow(terrain);
```

Boundaries and Movement

The *path finder* module from Chapter 4 cannot be reused directly. Currently, altitude is encoded as an integer from range [0, 255] and we may expect many points to share altitude. Consequently, we must also consult the slope to decide about the lowest neighbor. Therefore, there is a need to more precisely describe the *lowest neighbor* concept as follows:

- An adjacent point with the lowest altitude is the lowest neighbor.

- An adjacent point with the same altitude (if the others are at higher altitude) is the lowest neighbor, if the current point's slope is greater than zero.

- If there are multiple adjacent points satisfying the previous condition, then one of them is chosen in clockwise direction starting at south-west.

We obviously don't need the blue color component (aspect). The next line of code sets this layer to zero:

```
terrain[:, :, 2] = 0
```

Path Finding Engine

The original version of the path finding engine was implemented as a set of routines. To enhance maintainability, we will re-engineer the code to use object-oriented constructs. Listing 5-1 shows the backbone structure. This is the content of the base_patfinder.py file (it passes the Spyder's static code analyzer with maximum score 10/10). Notice how the embedded documentation for abstract methods doesn't mention anything about specific implementation details. These should be left to child classes.

Listing 5-1. Restructured Path Finding Module, Which Uses Object-Oriented Programming

```
"""The base class for implementing various path finders."""

import abc
from typing import List, Tuple, Set

import numpy as np

class BasePathFinder(metaclass=abc.ABCMeta):
    """

    Finds the path of a ball that descends in a terrain from
    some starting
    position.

    Args:
    terrain: the terrain's configuration comprised from
    (altitude, slope)
    integer pairs.
    """

    def __init__(self, terrain: np.ndarray):
        self._terrain = terrain

    @property
    def terrain(self):
        """Gets the current terrain data."""
        return self._terrain

    def wall(self, position: Tuple[int, int]) -> bool:
        """

        Checks whether the provided position is hitting
        the wall.
```

Args:
position: the pair of integers representing the ball's potential position.

Output:
True if the position is hitting the wall, or False otherwise.

Examples:
```
>>> BasePathFinder.__abstractmethods__ = set()
>>> path_finder = BasePathFinder(np.array([[(-2, 0),
(3, 0), (2, 0), (1, 0)]]))
>>> path_finder.wall((0, 1))
False
>>> BasePathFinder.__abstractmethods__ = set()
>>> path_finder = BasePathFinder(np.array([[(-2, 0),
(3, 0), (2, 0), (1, 0)]]))
>>> path_finder.wall((-1, 0))
True
"""

curr_x, curr_y = position
length, width = self.terrain.shape[:2]
return (curr_x < 0) or (curr_y < 0) or (curr_x >=
length) or (curr_y >= width)
```

@abc.abstractmethod
```
def next_neighbor(self, position: Tuple[int, int],
                visited: Set[Tuple[int, int]]) ->
                Tuple[int, int]:
    """

    Returns the position of the lowest neighbor or the
    current position.
```

```
        Args:
        position: the pair of integers representing the ball's
        current position.
        visited: the set of visited points.

        Output:
        The position (pair of coordinates) of the lowest
        neighbor.
        """

    @abc.abstractmethod
    def find_path(self, position: Tuple[int, int],
                visited: Set[Tuple[int, int]]) ->
                List[Tuple[int, int]]:
        """

        Finds the path that the ball would follow while
        descending in the terrain.

        Args:
        position: the pair of integers representing the ball's
        current position.
        visited: the set of visited points (may be preset to
        avoid certain points).

        Output:
        The list of coordinates of the path.
        """

if __name__ == "__main__":
    import doctest
    doctest.testmod()
```

Listing 5-2 shows our first attempt to implement the logic.

Listing 5-2. The Simple Path Finder Class That Uses Recursion to Traverse the Terrain

```
"""Simple recursive path finder implementation."""

from typing import List, Tuple, Set
from pathfinder.base_pathfinder import BasePathFinder

class SimplePathFinder(BasePathFinder):
    """Concrete path finder that uses recursion and is
    sequential."""

    def next_neighbor(self, position: Tuple[int, int],
                      visited: Set[Tuple[int, int]]) ->
                      Tuple[int, int]:
        """
        Uses a simple clockwise search of neighbors starting at
        south-west.

        Example:
        >>> path_finder = SimplePathFinder(np.array([[(-2, 0),
        (3, 0), (2, 0), (1, 0)]]))
        >>> path_finder.next_neighbor((0, 1), set((0, 1)))
        (0, 0)
        """

        curr_x, curr_y = position
        current_slope = self.terrain[position][1]
        min_altitude = self.terrain[position][0]
        min_position = position
```

```
        for delta_x in range(-1, 2):
            for delta_y in range(-1, 2):
                new_position = (curr_x + delta_x, curr_y +
                delta_y)
                if not self.wall(new_position) and not new_
                position in visited:
                    new_altitude = self.terrain[new_position][0]
                    if new_altitude < min_altitude or (new_
                    altitude == min_altitude and
                                            current_slope > 0):
                        min_altitude = new_altitude
                        min_position = new_position
        return min_position

    def find_path(self, position: Tuple[int, int],
                  visited: Set[Tuple[int, int]] = None) ->
                  List[Tuple[int, int]]:
        """

        Recursively finds the path.

        Example:
        >>> path_finder = SimplePathFinder(np.array([[(-1, 2),
        (-2, 1), (-2, 2), (1, 0)]]))
        >>> path_finder.find_path((0, 2))
        [(0, 2), (0, 1)]
        """

        if visited is None:
            visited = set()
        visited.add(position)
        next_position = self.next_neighbor(position, visited)
```

```
    if position == next_position:
        return [position]
    return [position] + self.find_path(next_position,
    visited)

if __name__ == "__main__":
    import doctest
    doctest.testmod()
```

Even this simple path finder is considerably more complex than our initial variant from Chapter 4. It uses a multidimensional Numpy array instead of a matrix for representing terrain data, has a different logic inside the next_neighbor method, and traces already visited points. Without this last step it could easily trigger an infinite recursion. The next code cell makes a simple test whose output is also shown (marked with >>>):

```
# Test the simulator with a toy terrain; slope is set to zero,
so only altitude matters.
test_terrain = np.array(
    [[(-2, 0), (3, 0),  (2, 0),  (1, 0)],
     [(-2, 0), (4, 0),  (3, 0),  (0, 0)],
     [(-3, 0), (3, 0),  (1, 0), (-3, 0)],
     [(-4, 0), (2, 0), (-1, 0),  (1, 0)],
     [(-5, 0), (-7, 0),  (3, 0),  (0, 0)]])

from pathfinder import SimplePathFinder
path_finder = SimplePathFinder(test_terrain)
path_finder.find_path((1, 1))
>>> [(1, 1), (2, 0), (3, 0), (4, 1)]
```

All seems to be OK, so we can try it out on our real terrain data. The following snippet of code is reused from the notebook presented in Chapter 4:

```
path_finder = SimplePathFinder(terrain)
interact(lambda start_x, start_y: path_finder.find_
path((start_x, start_y)),
        start_x = widgets.IntSlider(value=1, max=terrain.
        shape[0]-1, description='Start X'),
        start_y = widgets.IntSlider(value=1, max=terrain.
        shape[1]-1, description='Start Y'));
```

By playing around with the sliders, you may watch as paths get calculated. For example, a starting position (2966, 1367) would result in the following path:

```
[(2966, 1367), (2967, 1368), (2968, 1369), (2969, 1370),
(2970, 1371), (2970, 1372), (2971, 1373), (2972, 1374),
(2973, 1375), (2974, 1375), (2975, 1375), (2976, 1375),
(2977, 1375)]
```

Retrospective of Version 1.1

Apparently, the input/output subsystems are inapt. Setting a starting position with sliders and getting back a list of coordinates is cumbersome. Sure, it was OK for a toy terrain, but now the situation is different. Where is the point (2966, 1367) located on the map? How do we understand the ball's movement? Figure 5-2 shows the radar diagram that depicts the level of sophistication achieved across various dimensions in the initial version, and Figure 5-3 shows the level of sophistication achieved in this version.

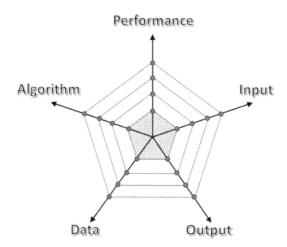

Figure 5-2. *The shaded area represents the overall progress across all dimensions; dimensions may be arbitrarily chosen depending on the context*

A system may attain different levels between them, as you will soon see. Our initial version has everything at the basic rank. The data dimension encompasses multiple aspects: volume, velocity, structural complexity, etc. The algorithm dimension represents both mathematical models and computer science algorithms. Input and output denote the capabilities pertaining to data acquisition and presentation of results, respectively. Performance in our case is associated with memory and CPU consumption.

171

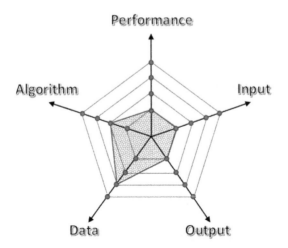

Figure 5-3. *The new version has moved two units on the data axis (bigger size and enhanced structure) and one on the algorithm line*

We have already seen that the input and output subsystems are deficient, so these require an improvement.

Retrospectives are regularly used in various Agile methods, too. They create opportunities for discussing lessons learned and potential improvements. Data science endeavors also require retrospectives. As you will see in the upcoming iterations, many times you will discover that there is a need for more exploration, preprocessing, and data-gathering cycles. The whole process is iterative and incremental.

Version 1.2

In this version we will move both the input and output subsystems by one unit; open the `Terrain_Simulation_v1.2.ipubn` notebook in the `augmented_ball_descend` source folder of this chapter. The idea is to allow the user to enter the starting position by directly specifying a point on the map. Furthermore, we would like to visualize the output on the map instead of just returning a set of points.

Enhancing the Input Subsystem

Let's start with Listing 5-3, which shows a small HTML CSS to customize the interactive matplotlib session (matplotlib is discussed in depth in Chapter 6). Try skipping this code cell to see its effect.

Listing 5-3. Removes the Default Title and Button for Closing the Interactive Session

```
%%html
<!-- Removes the default title "Figure 1" and button for an
interactive session. -->
<style>
    .output_wrapper .ui-dialog-titlebar {display: None;}
</style>
```

Listing 5-4 presents the main section in the notebook to gather input from a user. It uses an auxiliary class that is listed in Listing 5-5 (see also Exercise 5-2 at the end of the chapter). It differentiates between clicking a mouse to select an area of an image and selecting the starting position.

Listing 5-4. Setup of an Interactive Session to Acquire the Starting Position from a User

```
%matplotlib notebook

from ipywidgets import Textarea
from interactionlib import InteractionMonitor

fig = plt.figure()
plt.imshow(terrain)
```

```
info_area = Textarea(
    value = ",
    placeholder = 'Select a point on the map by clicking the
    mouse.',
    description = 'Position:',
    disabled = True
)
display(info_area)

fig.tight_layout(pad = 1.3)
interaction_monitor = InteractionMonitor(fig, info_area)
interaction_monitor.start()
```

Listing 5-5. Helper Class to Monitor User Actions

```
"""
Monitors whether the user is selecting an area on the image or
has chosen the
starting position.
"""

from ipywidgets import Textarea
import matplotlib.pyplot as plt

class InteractionMonitor:
    """
    Detects mouse events to figure out what a user is doing.

    Args:
    fig: the matplotlib figure to monitor.
    info_area: the external informational area whose value
    needs to be updated.
```

```
auto_stop_interaction: should interaction stop (when True)
after selecting
the starting position or not.
"""

def __init__(self, fig: plt.Figure, info_area: Textarea,
             auto_stop_interaction: bool = True):
    self._fig = fig
    self._info_area = info_area
    self._auto_stop_interaction = auto_stop_interaction
    self._cids = None
    self._selecting = False
    self._clicked = False
    self._clicked_position = None

def _on_click(self, event):
    self._clicked = True

def _on_release(self, event):
    if not self._selecting:
        self._clicked_position = (int(event.xdata),
        int(event.ydata))
        self._info_area.value = str(self._clicked_position)
        if self._auto_stop_interaction:
            self.stop()

    self._selecting = False
    self._clicked = False

def _on_motion(self, event):
    self._selecting = self._clicked
```

```
@property
def clicked_position(self):
    """Returns the clicked data position on the map."""
    return self._clicked_position

def start(self):
    """Starts monitoring mouse events on figure."""
    self._cids = [
        self._fig.canvas.mpl_connect('button_press_event',
        self._on_click),
        self._fig.canvas.mpl_connect('button_release_
        event', self._on_release),
        self._fig.canvas.mpl_connect('motion_notify_event',
        self._on_motion)]

def stop(self):
    """Closes the figure and stops the interaction."""
    plt.close(self._fig)
```

Caution   The previously implemented matplotlib interactive session only works with Jupyter Notebook (at the time of this writing). If you try to run the last code cell inside JupyterLab, you will receive the error `Javascript Error: IPython is not defined`. This is a fine example of the fact that you must occasionally solve problems completely unrelated to your main objectives.

Figure 5-4 illustrates a session of choosing the starting position. Now it is obvious where the ball will start to descend. All in all, this concludes the input part, although myriad additional enhancements are possible (for example, allowing users to directly enter coordinates, enabling navigation via keyboard, etc.).

pan/zoom, x=2398.54 y=1104.04 [196, 28, 0]

Position: Select a point on the map by
           clicking the mouse.

Figure 5-4. *Part of the original image zoomed by a user to easily select a desired point (second button from the right)*

You can see that the coordinates on axes reflect this zooming. The actual cursor position is printed on the right side together with data about the current altitude, slope, and aspect (this component is zeroed out so the whole picture is a combination of red and green). The small triangle above the coordinates is for resizing the whole UI. Once a point is selected, the session closes and its coordinates are displayed in the text area on the bottom. Compare this UI with the previous rudimentary slider-based input mechanism.

Enhancing the Output Subsystem

Handling the output will be an easier task. The system just needs to encode the path of a ball using the image's blue color component (it will be set to 255). This new image may be reused for another run. Accordingly, the notebook will save the final image for later use. Of course, the raw coordinates may still be valuable, so the system should return those, too.

One possible way to achieve the preceding goals is to have a new utility method that will visualize the calculated path on the terrain. The separation-of-concerns principle suggests that we should put this method into a new class, PathUtils. It will require the terrain and path as arguments and will return a new image. Listing 5-6 shows the implementation details of this method, while Listing 5-7 reveals the code cell that manages the workflow. This concludes the work on extending the output subsystem.

Listing 5-6. Implementation of the PathUtils Class, Currently Containing One Finalized Static Method

```
"""Contains various path related utility classes and
methods."""

from typing import List, Tuple

import numpy as np

class PathUtils:
    """Encompasses static methods to handle paths."""

    @staticmethod
    def encode_path(terrain: np.ndarray, descend_path:
    List[Tuple[int, int]]) -> np.ndarray:
        """
```

Encodes the path into the terrain by setting the points' 3rd (blue) component to 255.

Args:
terrain: the terrain's configuration comprised from (altitude, slope, [aspect])
integer pairs/triples.

Output:
New terrain with an extra 3rd dimension to encode the path.

Example:
```
>>> terrain = np.array([[(-1, 2), (-2, 1), (-2, 2), (1, 0)]])
>>> PathUtils.encode_path(terrain, [(0, 2), (0, 1)])
array([[[ -1,    2,    0],
        [ -2,    1, 255],
        [ -2,    2, 255],
        [  1,    0,    0]]])
"""
```

```python
# Expand terrain with an extra dimension, as needed.
if terrain.shape[2] == 2:
    new_shape = terrain.shape[:2] + (3,)
    new_terrain = np.zeros(new_shape, terrain.dtype)
    new_terrain[:terrain.shape[0], :terrain.shape[1],
    :2] = terrain
  else:
    new_terrain = np.copy(terrain)

for point in descend_path:
    new_terrain[point][2] = 255
return new_terrain
```

```
    @staticmethod
    def decode_path(terrain: np.ndarray) -> List[Tuple[int, int]]:
        """

        Decodes the path from the terrain by picking points
        whose 3rd (blue) component is 255.
        The reconstructed path may not be unique, which depends
        upon the path finder logic.

        Args:
        terrain: the terrain's configuration encoded with a
        single path.

        Output:
        The decoded path that is guaranteed to contain all
        points of the encoded path.
        Ordering of points may differ from what was reported by
        the matching path finder.
        """

        # Extra exercise to implement this method according to
        the specification.
        raise NotImplementedError

if __name__ == "__main__":
    import doctest
    doctest.testmod()
```

Listing 5-7. Notebook Cell That Uses the New Path Encoder Facility for Presenting Results

```
path_finder = SimplePathFinder(terrain)
x, y = interaction_monitor.clicked_position
calculated_path = path_finder.find_path((x, y))
print(calculated_path)
```

```
%matplotlib inline
from pathfinder import PathUtils

terrain = PathUtils.encode_path(terrain, calculated_path)
plt.figure(figsize=(13, 15))
# Zoom into area around the calculated path.
plt.imshow(terrain[max(x - 100, 0):min(x + 100, terrain.
shape[0]),
                    max(y - 100, 0):min(y + 100, terrain.
                    shape[1])]);

new_image_path = 'terrain_data/coastline_with_path.jpg'
imageio.imwrite(new_image_path, terrain)
```

Retrospective of Version 1.2

A lot of effort has been invested into enhancing the user experience (both to gather input and to present a result). UI work commonly requires a considerable amount of time and care. Moreover, it demands highly specialized professionals, which again reminds us to look at data science as a discipline that requires team work. Neglecting this fact may lead to creating an unpopular data product. For example, it is a well-known fact in building recommender systems that user experience is as important as the fancy back-end system. Figure 5-5 shows the radar diagram for this version.

If you try out the current revision, you will notice that most calculated paths are very short. This probably comes as a surprise; after all, the terrain data is quite large. There are several reasons and possible solutions to produce more realistic outputs, as listed here:

- The input terrain data is coarsely grained, where 256 levels of altitude isn't enough. Observe that our path finder can accept a higher-fidelity configuration without any change. So, one option is to look for images with higher color depth or abandon images altogether and rely on raw data.

- The current algorithm to find neighbors is rudimentary. We all know from physics that a ball with high momentum can also roll up a slope. Furthermore, it will roll for some time even on a flat surface. Therefore, we may implement a new path finder that would better model the movement of a ball. One simple approach would be just to introduce a new state variable momentum. The ball's momentum would increase at each spot with a positive slope toward a lower altitude, and its momentum would decrease at each spot after the lowest altitude is reached. On a flat surface, a ball would lose some small amount of its momentum at each iteration. Try to implement this as an additional exercise and see what happens.

- The constraint of disallowing a ball to bounce back to a visited spot may also be a limiting factor. With an advanced model, you may allow such situations, especially if you introduce some randomness, too.

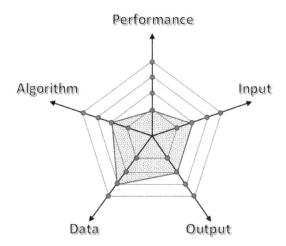

Figure 5-5. *You should expect to need to improve performance soon after the publication of a useful, polished, and powerful data product. It cannot stay too long at its basic level.*

Software architecture is the chief element for delivering nonfunctional requirements, including performance. This is why you must take care to structure your solution in a maintainable and evolvable manner with well-defined and connected components. Any optimization should be done on major pieces of the system, where the return on investment in performance tuning is most beneficial; measure performance (possibly with a profiler), consult the performance budget dictated by an architecture, and finally start making changes. If components are loosely coupled and expose clear APIs, you may be safe against ripple effects. Of course, covering your code with automated tests is something you should do by default. All these quality aspects will help you continue with rapid explorations, something you expect to be capable of by using Python and Jupyter Notebook technology. The secret is that your code base must also be in proper shape for magic to happen. Just think about the mess that would ensue if everything were piled up inside a single notebook, without information hiding and encapsulation (these were attained by using modules and object-oriented constructs).

There are two major reworks for the next version (presume that the current version has already attracted a huge number of diverse users who are accessing our service at a high rate):

- Allow handling of arbitrarily large terrains. The current solution imposes an artificial limit on path size (the depth of recursion depends on stack size).

- Allow calculating multiple paths at once (we will also need to update the UI to support this). We should also improve performance by using all available local and/ or distributed hardware resources.

Version 1.3

In general, there are three approaches to adding novelties and technology upgrades to your system:

1. Implement your own proprietary solutions without reusing available open-source or commercial artifacts.

2. Mostly reuse existing frameworks and spend time only on gluing things together.

3. Combine the previous two options.

You would go with the first choice if you could not find a suitable framework to reuse. There is little chance that this will be the case in most situations. Therefore, it is always beneficial before any work to look around and see what is offered. Remember that any homemade software demands expertise to develop and fully maintain. The former is frequently a huge barrier to most developers and organizations. For example, a golden rule in security is to never invent your own encryption algorithm. Publicly exposing software is a potent way to ensure quality, since many users would report all sorts of issues and try it out in a multitude of scenarios.

At the opposite end of the spectrum is the eager reuse of software solutions. This option is again troublesome for several reasons. It is difficult to find for any complex problem an out-of-the-box solution. Even if you find one, it is usually arduous to overcome the limitations imposed by reused software; jumping out from a presupposed usage model is very costly. So, you must be confident that the model favoring a reused system will be right for you for a long period of time.

So, we come to the third choice, which balances between development and reuse. This alternative is the default one for most data scientists. Nonetheless, the *reuse* part is not without dangers, and I'm not referring here solely to security holes in open-source software. It makes no sense to invest energy and time learning a new framework only to soon abandon it because it isn't appropriate. Hectic reuse may cost even more than doing everything from scratch; don't forget that changing a core framework (impacting an architecture) may trigger a rewrite of a large amount of existing code. You must be aware of the following characteristics of reuse-based software engineering:

- Often you will need to relax or alter the project's initial set of requirements when considering a particular framework. If you don't have such requirements, then you should first produce a proper project charter and gather such requirements. Consequently, be ready to revisit requirements with all pertinent stakeholders on the project. It isn't uncommon in large organizations that requirements get ignored just because, somewhere down the road, a developer isn't able to force a framework into a particular usage scenario.

- When choosing frameworks, always consider many candidates. Each has its own strengths and weaknesses, so you will have to make a compromise based on the core objectives of your data product.

- Every new framework requires some amount of familiarization and training. You also must ensure that your teammates are acquainted with the framework.

- Always check the freshness of the framework and what kind of support you can get. There is nothing worse than bringing an obsolete library into your new product.

- Every framework has a considerable impact on your design. Therefore, ensure that a later switch would not create a chain reaction (research the dependency inversion principle online).

Establishing the Baseline

To monitor progress, we need to specify the baseline. The current system has difficulties evaluating large terrains with long paths. For performance-testing purposes, we will create a degenerate terrain configuration to expose various running times (worst, average, and best case). We will produce a square surface whose spots are sorted according to altitude in ascending order and then augmented with "walls" of maximum altitude (like a maze). The worst-case performance can be measured by always starting from a lower-right corner. The best case would hit the top-left point. Randomly picking points and taking their average processing time would designate the average case. We can also make a terrain large enough to blow up the memory constraint (i.e., not being able to store it in memory).

We will work solely inside Spyder using its built-in console for investigation. You may want to record all steps by creating automated tests. The function to create our test terrain is shown in Listing 5-8 (located in the `testutils/create_terrain.py` file).

Listing 5-8. Function to Produce a Maze-like Test Terrain (with Embedded Example)

```python
def create_test_terrain(n: int) -> np.ndarray:
    """Creates a square maze-like terrain with alleys of
    decreasing altitude.

    Args:
    n: number of rows and columns of a terrain

    Output:
    The test terrain of proper size.

    Example:
    >>> terrain = create_test_terrain(9)
    >>> terrain[:, :, 0]
    array([[ 0,  1,  2,  3,  4,  5,  6,  7,  8],
           [81, 81, 81, 81, 81, 81, 81, 81, 17],
           [26, 25, 24, 23, 22, 21, 20, 19, 18],
           [27, 81, 81, 81, 81, 81, 81, 81, 81],
           [36, 37, 38, 39, 40, 41, 42, 43, 44],
           [81, 81, 81, 81, 81, 81, 81, 81, 53],
           [62, 61, 60, 59, 58, 57, 56, 55, 54],
           [63, 81, 81, 81, 81, 81, 81, 81, 81],
           [72, 73, 74, 75, 76, 77, 78, 79, 80]])
    """

    size = n * n
    terrain = np.zeros((n, n, 2), dtype=int)
    terrain[:, :, 0] = np.arange(0, size).reshape((n, n))

    # Reverse every 4th row to have proper ordering of
    elements.
    for i in range(2, n, 4):
        terrain[i, :, 0] = np.flip(terrain[i, :, 0])
```

```
# Create "walls" inside the terrain.
for i in range(1, n, 4):
    terrain[i, :-1, 0] = size
for i in range(3, n, 4):
    terrain[i, 1:, 0] = size

return terrain
```

Listing 5-9 shows the extra find_paths method that is added to the BasePathFinder class. Observe that all child classes will immediately inherit this capability. This is a perfect example of how proper object orientation can boost your efficacy.

Listing 5-9. Method That Calls find_path for Each Starting Position

```
def find_paths(self, positions: List[Tuple[int, int]]) ->
List[List[Tuple[int, int]]]:
    """
    Finds paths for all provided starting positions.

    Args:
    positions: the list of positions to for which to calculate
    path.

    Output:
     The list of paths in the same order as positions.
    """

    return [self.find_path(position, None) for position in
    positions]
```

The following snippet makes some test runs (type each statement in Spyder's console, and make sure you are in the augmented_ball_descend subfolder of this chapter's source code):

```
>> from testutils import create_test_terrain
>> from pathfinder import SimplePathFinder
>>
>> test_terrain = create_test_terrain(10001)
>> path_finder = SimplePathFinder(test_terrain)
>>
>> %time calc_path = path_finder.find_path((0, 0))
CPU times: user 53 µs, sys: 39 µs, total: 92 µs
Wall time: 105 µs
>>
>> %time calc_paths = path_finder.find_paths([(0, 0), (0, 2000)])
CPU times: user 44.6 ms, sys: 1.75 ms, total: 46.3 ms
Wall time: 45.1 ms
>>
>> %time calc_path = path_finder.find_path((0, 3000))
...lines omitted...
RecursionError: maximum recursion depth exceeded
```

OK, we cannot even go above 3000 points. Let's get rid of recursion and continue our test runs. Listing 5-10 shows a nonrecursive version of the simple path finder class. This revision cannot match the elegance of the recursive idea, but we cannot chose here the most natural problem-solving paradigm. The story would be different in pure functional languages.

Listing 5-10. Nonrecursive Simple Path Finder That Will, at Least Theoretically, Be Able to Process All Starting Positions

```
"""Simple nonrecursive path finder implementation."""

from typing import List, Tuple, Set
from pathfinder.simple_pathfinder import SimplePathFinder

class NonRecursiveSimplePathFinder(SimplePathFinder):
    """Concrete path finder that doesn't use recursion."""

    def find_path(self, position: Tuple[int, int],
                  visited: Set[Tuple[int, int]] = None) ->
                  List[Tuple[int, int]]:
        """
        Iteratively finds the path (without using recursion).

        Example:
        >>> path_finder = NonRecursiveSimplePathFinder(np.
        array([[(-1, 2), (-2, 1), (-2, 2),➡
(1, 0)]]))
        >>> path_finder.find_path((0, 2))
        [(0, 2), (0, 1)]
        """

        if visited is None:
            visited = set()
        visited.add(position)
        calculated_path = [position]
        next_position = self.next_neighbor(position, visited)
```

```
    while position != next_position:
        position = next_position
        visited.add(position)
        calculated_path.append(position)
        next_position = self.next_neighbor(position, visited)

    return calculated_path

if __name__ == "__main__":
    import doctest
    doctest.testmod()
```

The following statements exemplify this new class:

```
>> from pathfinder import NonRecursiveSimplePathFinder
>> path_finder = NonRecursiveSimplePathFinder(test_terrain)
>>
>> %time calc_paths = path_finder.find_paths([(0, 0),
(1000, 1000)])
CPU times: user 1min 46s, sys: 830 ms, total: 1min 46s
Wall time: 1min 46s
```

The nonrecursive version is capable of handling much longer paths, which shouldn't surprise us when programming in Python (the recursive variant relies on *tail recursion*, which isn't expensive in pure functional languages). Nevertheless, the running time of getting from point (1000, 1000) to the lowest point is already exorbitantly slow. There is no need to try with longer distances.

Performance Optimization

If we just announce a vague aim that "the code must run faster" before actual work, then we would probably fail in our optimization attempt. We must meet all of the following conditions:

- The bottleneck and critical parts of the code are identified; running your code under a profiler is mandatory.

- The use cases that must be improved are evident; this defines the overall context of optimization. People often optimize for scenarios that will never occur in production.

- The structure of your code base is of adequate quality and covered with tests to avoid regression and chaos.

Figure 5-6 shows a typical sequence of action between a user and our system; we will only focus on a specific user, but keep in mind that our service needs to support multiple users in parallel. From this use case we may conclude that many simulations will happen over the same terrain. All of those runs could execute in parallel, since the model is static (terrain data isn't altered in any way). The story would be different for a dynamic setup.

The following is output from a profiling session to calculate the multitude of paths at once:

```
>> %prun calc_paths = path_finder.find_paths([(0, 0), (0, 10), (0, 100), (0, 1000), ➥
(2, 10), (1000, 1000)])
        15059334 function calls in 145.934 seconds

Ordered by: internal time

ncalls   tottime  percall  cumtime  percall  filename:lineno(function)
5022106  84.972   0.000    136.366  0.000    simple_pathfinder.py:9(next_neighbor)
45198954 39.729   0.000    45.899   0.000    base_pathfinder.py:26(wall)
90333057 11.665   0.000    11.665   0.000    base_pathfinder.py:21(terrain)
      6   5.339   0.890    143.266  23.878   non_recursive_simple_pathfinder.py:9(find_path)
      1   2.505   2.505    145.933  145.933  <string>:1(<module>)
5022106   1.051   0.000    1.051    0.000    {method 'add' of 'set' objects}
5022100   0.510   0.000    0.510    0.000    {method 'append' of 'list' objects}
      1   0.162   0.162    143.428  143.428  base_pathfinder.py:90(<listcomp>)
      1   0.001   0.001    145.934  145.934  {built-in method builtins.exec}
      1   0.000   0.000    143.428  143.428  base_pathfinder.py:79(find_paths)
      1   0.000   0.000    0.000    0.000    {method 'disable' of '_lsprof.Profiler' objects}
```

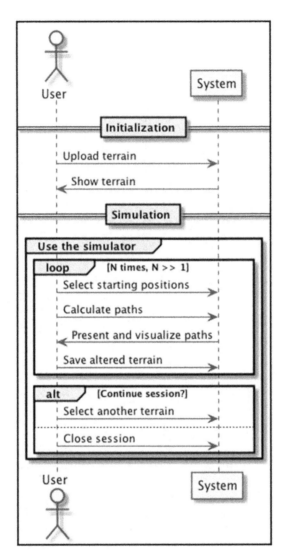

Figure 5-6. *This sequence of interaction will happen independently for each user*

We assume and optimize for a situation where the number of trials per terrain is large (>> 1). This may have a negative impact on extreme cases of 1 simulation per terrain. Again, this is a fine example of why you should be clear about your use cases.

There are three tactics to optimize a program:

- Eliminate or reduce the number of calls to methods.

- Speed up methods.

- Combine the previous two options.

Obviously, a tremendous number of calls target the `terrain` property and `wall` method. These calls solely happen from the `next_neighbor` method, which is called once for each point in a path. If calculating a next point would become computationally demanding in the future, then we could eliminate the need to "recalculate" the terrain over and over again by preprocessing it during initialization (see Figure 5-6). Afterward, we would just pick up the next points by interrogating the index. In some sense, you can think of this as a sort of *memoization*. Figure 5-7 depicts the idea visually. Of course, the drawback is extra memory usage to store preprocessed stuff. Clearly, doing this is only beneficial if there are many simulations over the same terrain and their real-time processing is expensive. Otherwise, it is a total waste of time and memory. We will only speed up the `next_neighbor` method and leave preprocessing as an additional exercise.

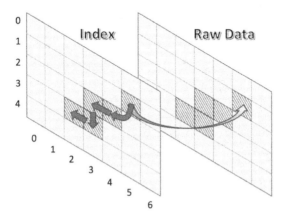

Figure 5-7. *Here, the raw terrain data isn't used at all during simulation*

The next_neighbor method simply reads out next points based upon the current position. Preprocessing is performed by going over every point and storing its best neighbor (it may also be the current point if the ball cannot go anywhere else). For example, the index cell (1, 5) references the next neighbor, whose value is (2, 4). Reading the raw terrain data at (1, 5) yields that point's altitude and slope.

The find_path method is called as many times as there are starting positions in the list. If we may parallelize find_path executions, then we may decrease the find_paths method's execution time.

All in all, we have a plan and we are ready for implementation. At this moment, we must rethink how to proceed with construction from the viewpoint of reuse. We definitely should rely on some mature framework, as the realm of parallel and distributed computing is abundant with offerings. Here are our requirements that we expect from the new framework:

- It must support both explicit and implicit parallelism. So, we are not looking solely for an abstraction over which we can reach out to many cores or machines. For example, Apache Spark exposes a straightforward model through the *resilient distributed dataset* abstraction but forces us to formulate problems in a specific way. For arbitrarily complex logic, we would need something more akin to generic frameworks, like *multiprocessing* in Python.

- We don't want to rewrite much of the existing code nor depart into another run-time environment. For example, a JVM-based technology isn't preferable.

- Preferably, we would like to retain the current APIs when working with new abstractions. This would also reduce learning time.

It turns out that there is an open-source framework that meets our aspirations. Numba (see `http://numba.pydata.org`) is a just-in-time (JIT) compiler that translates Python functions with Numpy code into machine code. There is no special precompilation phase. All you need to do is annotate functions that you wish to accelerate. Numba integrates seamlessly with Dask and may even produce universal functions that behave like built-in Numpy functions. Being able to speed up arbitrary functions is very important when you have complex business logic with lots of conditionals. In this case, any crude parallelization scheme would become cumbersome to apply. Finally, with Numba you can optimize for GPUs, again something very crucial to gain top performance.

Listing 5-11 shows the new `ParallelSimplePathFinder` class. It uses the logic of the inherited `next_neighhbor` method to create an index, so the ball movement rules remain intact. All that was required to boost performance is that single decorator. Of course, you need to know the constraints regarding content of `jit` decorated functions. This is why you should always keep `nopython=True` set (otherwise you will not get warnings).

Listing 5-11. Performance-Tuned Path Finder Class That Uses Numba

```python
"""Efficient parallel version of the path finder system."""

from typing import Tuple, Set

import numpy as np
from numba import jit
from pathfinder.non_recursive_simple_pathfinder import
NonRecursiveSimplePathFinder

class ParallelSimplePathFinder(NonRecursiveSimplePathFinder):
    """Concrete path finder that uses Numba to perform
    operations in parallel."""
```

```python
    @staticmethod
    @jit(nopython=True, parallel=True, cache=True)
    def _best_neighbor(terrain: np.ndarray, position:
    Tuple[int, int]) -> Tuple[int, int]:
        curr_x, curr_y = position
        length, width = terrain.shape[:2]
        current_slope = terrain[position][1]
        min_altitude = terrain[position][0]
        min_position = position

        for delta_x in range(-1, 2):
            for delta_y in range(-1, 2):
                new_position = (curr_x + delta_x, curr_y +
                delta_y)
                new_x, new_y = new_position
                if not ((new_x < 0) or
                        (new_y < 0) or
                        (new_x >= length) or
                        (new_y >= width)) and not new_position
                        == position:
                    new_altitude = terrain[new_position][0]
                    if new_altitude < min_altitude or (new_
                    altitude == min_altitude and
                                            current_slope > 0):
                        min_altitude = new_altitude
                        min_position = new_position
        return min_position

    def next_neighbor(self, position: Tuple[int, int],
                    visited: Set[Tuple[int, int]]) ->
                    Tuple[int, int]:
        """
```

Uses a vectorized clockwise search of neighbors starting at south-west.

Example:
```
>>> terrain = np.array([[(-2, 0), (2, 0), (2, 1),
(3, 1)]])
>>> path_finder = ParallelSimplePathFinder(terrain)
>>> path_finder.next_neighbor((0, 2), set((0, 2)))
(0, 1)
"""

        best_neighbor = ParallelSimplePathFinder._best_
        neighbor(self.terrain, position)
        if not best_neighbor in visited:
            return best_neighbor
        return position

if __name__ == "__main__":
    import doctest
    doctest.testmod()
```

Let's now repeat the command that we used in the baseline version (the path_finder variable points to our new class):

```
>> %time calc_paths = path_finder.find_paths([(0, 0),
(1000, 1000)])
CPU times: user 8.42 s, sys: 395 ms, total: 8.82 s
Wall time: 8.81 s
```

So, execution time went down from 1min 46s to 8.82s! This is more than a 10× improvement. The good news is that there are tons of more possibilities to decrease the running time (indexing, mentioned earlier, is one of them).

Retrospective of Version 1.3

This version was intended to show you how a properly structured code base helps evolution. We made small modifications in each increment, without spending time to fight clutter. The Python world is abundant with frameworks that wait to be reused. You must avoid the danger of being unduly influenced by any of them. To ensure that your program is always amenable to additional enhancements, you must control and assess what it requires to operate efficiently and avoid adding functionality that has dependencies. This concludes our series of iterations regarding this project. Figure 5-8 shows the radar diagram for this last version.

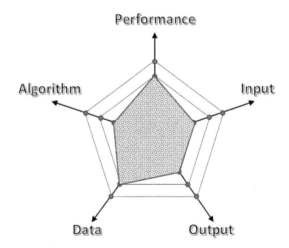

Figure 5-8. *The performance is now satisfactory, which allows us to add new, complex features*

Abstractions vs. Latent Features

Thus far, we have worked with software abstractions: frameworks, classes, data structures, etc. Abstractions are our principal mechanism to combat complexity. Working with entities on proper abstraction levels increases both efficacy and quality. These software abstractions have all been explicit, because we created them. In this section you will see that there is a totally different category of abstractions that emerges behind the scenes. We don't even know what they represent, although we know everything about the process responsible for their creation. Often these artifacts are called *latent features*.

Latent features describe raw data in a more concise fashion; by using them, we can achieve high compression in many aspects. Besides dropping the sheer data size, more importantly, we can reduce the number of dimensions in the dataset. This directly brings down complexity, as many machine-learning algorithms work much better in lower dimensional space (we will talk about machine learning in Chapter 7).

There are two fundamental compression procedures: fully explainable and partially explainable. In Chapter 2 you witnessed the former, where the application of mathematical statistics produced a more compact data representation. In other words, we summarized the dataset while preserving all pertinent details regarding customer behavior. In this section, you will see an example of partially explainable compression procedures.

Interpretability of your solution is a very important topic nowadays (see reference [1] for a good recount). For example, many companies want to know how and why particular items were recommended to users by a recommender system. You cannot just respond that it was decided by a machine that leverages a well-known matrix factorization technique. Remember this side-effect of latent features when you start to design a new system.

Compressing the Ratings Matrix

Suppose that you are building a recommender system and store ratings of users for items inside matrix R. Each row represents one user and each column represents an item. For user u and item i you can easily find out the rating r_{ui} by looking at the corresponding cell of the matrix. Obviously, may cells will be empty, because users will rate only a miniscule number of items. With millions of users and items, this matrix may become enormous (even if you store it as a sparse matrix). Apparently, no commercial recommender system uses such raw matrices.

There is a theorem of linear algebra stating that any matrix R_{mxn} with rank k can be decomposed into three subordinate matrices U_{mxk}, S_{kxk}, and V_{kxn}, such that $R = USV^T$. The columns of U and V are pairwise orthogonal, and the matrix S is diagonal (its elements are called *singular values*). Now, the magic is related to choosing the number of latent features d < k, so that we can abandon most of the elements in U, S, and V. Of course, the goal is to still be able to approximately reconstruct the original matrix R. The aim is to efficiently calculate predictions of ratings for users and items that haven't been rated (we presume a multivalued rating scheme, such as number of stars).[1]

The following snippet from Spyder's console creates a random dense ratings matrix R, factorizes it into subordinate matrices, and demonstrates what happens in a lower dimensional case (when d < rank(R)):

```
>> m = 10, n = 6
>> np.random.seed(10)
>> a = np.random.randint(1, 5, size=(m, n))
>> a
array([[2, 2, 1, 4, 1, 2],
       [4, 1, 2, 2, 1, 2],
       [2, 3, 1, 2, 1, 3],
       [1, 3, 1, 4, 1, 1],
       [4, 4, 3, 1, 4, 3],
       [3, 2, 1, 1, 3, 2],
```

---

[1]Of course, for this we would first need to normalize ratings by some bias and convert missing values to zero. Later, we would need to denormalize the prediction using the same bias. Going into these details would create an unnecessary detour from our main topic of demonstrating latent features.

```
      [4, 3, 2, 2, 2, 4],
      [2, 3, 2, 1, 3, 3],
      [3, 2, 1, 2, 4, 2],
      [1, 4, 3, 2, 2, 3]]])
>>
>> np.linalg.matrix_rank(a)
6
>>
>> u, s, vt = np.linalg.svd(a, full_matrices=True)
>> u.shape, s.shape, vt.shape
((10, 10), (6,), (6, 6))
>>
>> s
array([18.30683743,  4.73046021,  3.60649612,  2.79133251,
  1.65155672, 0.97798699])
>>
>> np.allclose(a, np.dot(u[:, :n] * s, vt))
True
>>
>> d = 5
>> np.allclose(a, np.dot(u[:, :d] * s[:d], vt[:d]))
False
>>
>> np.allclose(a, np.dot(u[:, :d] * s[:d], vt[:d]), atol = 0.3)
True
>>
>> u[:1, :d]
array([[-0.26299616, -0.50400408,  0.2691205 ,  0.10395761,
  0.06633331]])
```

The singular values are ordered in descending order, and we kept the first five latent features. Sure, we no longer can faithfully reconstruct R in a lower dimensional space, but we may estimate its values quite correctly. Ratings anyhow contain noise, and our goal isn't to figure out the known values. The mission is to predict unknown ratings, and for this purpose an approximation of R is good enough. The last line above shows the reduced vector for user 1. It is expressed with five latent features that somehow designate the "taste" of users. Without further deep analysis, we have no inkling what aspect of "taste" they denote.

This example only scratches the surface of recommender systems, but I hope that you have grasped the key idea.[2] You will encounter latent features everywhere; the principal component analysis method and neural networks are all based upon them. In some sense, instead of doing feature engineering yourself, you leave this job to a machine.

EXERCISE 5-1. RANDOM PATH FINDER

The `SimplePathFinder` class visits potential neighbors in a predefined order. Consequently, it will generate the same path every time (assuming no change in the starting position). This doesn't quite mimic the real world and is also a boring behavior.

Create a new class `RandomSimplePathFinder` as a child of `SimplePathFinder` and override the `next_neighbor` method. It should randomly select the next point among equally good candidates, if there is more than one. Feel free to refactor the parent class(es), if you judge that would reduce duplication of code.

---

[2]You may want to try out LensKit (see `https://lenskit.org`) to research and/ or build your own recommender system. It provides you all the necessary infrastructure and algorithms to focus only on specific aspects of your engine.

EXERCISE 5-2. MORES REUSABLE INTERACTION MONITOR

The InteractionMonitor class accepts a text area to update its value property each time a user selects a new starting position. This was OK for our purpose but may be inadequate in other situations.

Improve this class by making it more reusable. One option is to receive a function (instead of that informational area) accepting one tuple argument that denotes the selected starting position. In this manner, users could have full control over the selection process. One good use case is selecting multiple points. These could be collected by your custom function until the desired number of points is reached. Afterward, your function would call the stop method on the InteractionMonitor instance.

Summary

This chapter has demonstrated how software engineering methods and concepts interplay with those related specifically to data science. You have witnessed how progression through multiple interrelated dimensions requires careful planning and realization. These dimensions were performance, input, output, data, and algorithms. Radar diagrams are very handy to keep you organized and to visually represent strengths and weaknesses of the current system. You should have implicitly picked up the main message: as soon as you start doing something advanced in any particular area, things get messy. This is the principal reason why methods that work at a small scale don't necessarily work in large-scale situations.

There are tons of frameworks that we haven't even mentioned, but all of them should be treated similarly to how we treated Numba in this chapter. The next two paragraphs briefly introduce a few notable frameworks, and in upcoming chapters you will encounter some very powerful frameworks, such as PyTorch for deep neural networks (in Chapter 12). We will also come back to Dask in Chapter 11.

Dask (see `https://dask.org`) is a library for parallel computing in Python (consult references [1] and [2] for a superb overview). Dask parallelizes the numeric Python ecosystem, including Numpy, Pandas, and Scikit-Learn. You can utilize both the implicit and explicit parallelization capabilities of Dask. Furthermore, Dask manages data that cannot fit into memory as well as scales computation to run inside a distributed cluster. Dask introduces new abstractions that mimic the APIs of the underlying entities from the SciPy ecosystem. For example, with dask arrays you can handle multiple smaller Numpy arrays in the background while using the familiar Numpy API. There are some gotchas you need to be aware of. For example, don't try to slice large dask arrays using the `vindex` method (implements fancy indexing over dask arrays). At any rate, combining Dask and Numba can give you a very powerful ensemble to scale your data science products.

Luigi (see `https://github.com/spotify/luigi`) is a framework to build complex pipelines of batch jobs with built-in Hadoop support (this represents that harmonizing power between various frameworks). On the other hand, Apache Nifi (see `https://nifi.apache.org`) is useful to produce scalable data pipelines. As you can see, these two frameworks attack different problems, although they may be orchestrated to work together depending on your needs.

References

1. Steven Strogatz, "One Giant Step for a Chess-Playing
    Machine," *New York Times*, `https://www.nytimes.`
    `com/2018/12/26/science/chess-artificial-`
    `intelligence.html`, Dec. 26, 2018.

2. Matthew Rocklin, "Scaling Python with Dask,"
    Anaconda webinar recorded May 30, 2018, available
    at https://know.anaconda.com/Scaling-Python-
    Dask-on-demand-registration.html.

3. Dask Tutorial," `https://github.com/dask/dask-`
    `tutorial`.

CHAPTER 6

Data Visualization

Visualization is a powerful method to gain more insight about data and underlying processes. It is extensively used during the exploration phase to attain a better sense of how data is structured. Visual presentation of results is also an effective way to convey the key findings; people usually grasp nice and informative diagrams much easier than they grasp facts embedded in convoluted tabular form. Of course, this doesn't mean that visualization should replace other approaches; the best effect is achieved by judiciously combining multiple ways to describe data and outcomes. The topic of visualization itself is very broad, so this chapter focuses on two aspects that are not commonly discussed elsewhere: how visualization may help in optimizing applications, and how to create dynamic dashboards for high-velocity data. You can find examples of other uses throughout this book.

Visualization should permeate data science processes and products. It cannot be treated as an isolated entity, let alone be reduced to fancy diagrams. In this respect, I frown upon any sort of rigid categorization and dislike embarking on templatizing what kind of visualization is appropriate in different scenarios. There are general guidelines about encoding of variables (see reference [1]) using various plot types, and these should be followed as much as possible. Nonetheless, the main message is that if visualization helps to make actionable insights from data, then it is useful.

© Ervin Varga 2019
E. Varga, *Practical Data Science with Python 3*,
https://doi.org/10.1007/978-1-4842-4859-1_6

Algorithms are the cornerstone of doing data science and dealing with Big Data. This is going to be demonstrated in the next case study. Algorithms specify the asymptotic behavior of your program (frequently denoted in Big-O notation), while technologies including parallel and distributed computing drive the constant factors. We will apply visualizations of various sorts to showcase these properties. It is important to emphasize that the notion of a *program* may symbolize many different types of computations. Computing features is one such type. A fast algorithm may enable your model to properly run in both the batch regime and online regime. For an example using interactive visualization to show the implementation of data structures and algorithms, you may want to visit the Data Structure Visualizations web site (see `https://www.cs.usfca.edu/~galles/visualization`). Another interesting site for you to explore is Python Tutor, which is related to visualizing the execution of Python programs (visit `https://pythontutor.com`). It also has links to similar sites for learning other popular languages, like JavaScript, C++, Java, etc.

Producing static graphs is totally different than trying to depict major characteristics of data in continual flux. This will be our topic in the second part of this chapter. Efficient dashboards allow real-time monitoring of processes by visualizing business metrics, so that operators may promptly act upon them.

Visualizing Temperature Data Case Study

The aim of this section is to illuminate the architecture of matplotlib, which is the most famous Python visualization framework (you should read the definitive guide at `http://aosabook.org/en/matplotlib.html`). To make the topic more comprehensible, we will use a sample publicly available dataset from NOAA's Global Historical Climatology Network - Daily (GHCN-Daily), Version 3. This sample dataset contains measurements from a single station for January 2010. It is available in the source code for this chapter, but you can download it directly from https://www1.ncdc.noaa. gov/pub/data/cdo/samples/GHCND_sample_csv.csv. For each day we will use the columns denoting the minimum and maximum temperature readings given in tenths of degree Celsius. The station's latitude and longitude are expressed in the correspondingly named columns.

Figure 6-1 shows the major components of the matplotlib ecosystem. Many sophisticated extensions are built on top of it, like the Seaborn library (you will see it in action in later chapters). You may wonder why Pandas is shown here. Well, it has its own visualization subsystem that directly relies on matplotlib. Matplotlib also comes with a simple scripting library called pyplot. This is used for quick solutions that are especially useful during exploratory data analysis.

Figure 6-1. *Some higher-level components that sit on top of matplotlib, which is further segregated into smaller units that you can reference*

Showing Stations on a Map

Listing 6-1 shows the function to produce a geographic map with data sources (in our case the sole temperature sensor). The code uses the mplleaflet Python library for converting a matplotlib plot into a Leaflet map (see https://leafletjs.com). The plot_stations.py module inside the temp_plots folder for showing our data source on a map. It uses matplotlib in scripting regime via pyplot.

Listing 6-1. Module for Showing Our Data Source on a Map.

```
import matplotlib.pyplot as plt
import mplleaflet

def plot_stations(longitudes, latitudes, embedded = False):
    if embedded:
        plt.figure(figsize = (8, 8))
    plt.scatter(longitudes, latitudes,
                c = 'b',
                marker = 'D',
                alpha = 0.7,
                s = 200)
    return mplleaflet.display() if embedded else mplleaflet.show()
```

The embedded parameter should be set to True to include the generated map inline in a Jupyter notebook. Listing 6-2 shows the first part of the driver code.

Listing 6-2. First Part of the temp_visualization_demo.py Module

```
import pandas as pd

df = pd.read_csv('GHCND_sample_csv.csv',
                 usecols = [3, 4, 5, 6, 7],
                 index_col = 2,
                 parse_dates = True,
                 infer_datetime_format = True)
df['TMIN'] = df['TMIN'] / 10
df['TMAX'] = df['TMAX'] / 10
print(df.head())
```

```
from plot_stations import plot_stations
plot_stations(df['LONGITUDE'].tolist()[0], df['LATITUDE'].
tolist()[0])
```

The following is printed as output when this code is executed:

```
           LATITUDE  LONGITUDE  TMAX   TMIN
DATE
2010-01-01  48.0355     -98.01 -17.8 -31.1
2010-01-02  48.0355     -98.01 -24.4 -32.2
2010-01-03  48.0355     -98.01 -19.4 -28.9
2010-01-04  48.0355     -98.01 -16.7 -20.0
2010-01-05  48.0355     -98.01 -13.3 -16.7
```

Figure 6-2 shows the generated interactive map with a blue diamond denoting our temperature sensor. You will also find the _map.html file in the current working directory. This is automatically created by mplleaflet.show() (it is possible to change the output file name).

Plotting Temperatures

We will now plot both the minimum and maximum temperatures as well as overlay points that designate extremely high (> 0) and low (< –30) temperatures in Celsius. To make the plot more useful, we will add an additional y axis for showing temperatures in Fahrenheit, too. This time the code relies on matplotlib and Pandas visualization and illustrates how to switch between different abstraction layers. Listing 6-3 shows the second part of the driver code that prepares the data as well as calls our custom plotting function, which is shown in Listing 6-4 and displayed in Figure 6-3. The solution is also reusable in other contexts with minor modifications (for example, changing the titles of axes, altering the tick marks, etc.).

213

Figure 6-2. *The interactive map with our temperature sensor (a quite remarkable result with just a couple of lines of code)*

Listing 6-3. Second Part of the temp_visualization_demo.py Module

```
min_temp = df['TMIN'].min()
max_temp = df['TMAX'].max()
print("\nMinimum temperature: %g\nMaximum temperature: %g\n" %
(min_temp, max_temp))

LIMIT_HIGH = 0
LIMIT_LOW = -30

extreme_high_temps = df['TMAX'][df['TMAX'] > LIMIT_HIGH]
extreme_low_temps = df['TMIN'][df['TMIN'] < LIMIT_LOW]

print('Extreme low temperatures\n', extreme_low_temps)
print('\nExtreme high temperatures\n', extreme_high_temps)

from plot_temps import plot_temps
plot_temps(df, min_temp, max_temp, extreme_low_temps, extreme_
high_temps)
```

The following is the textual report generated by the previous code:

```
Minimum temperature: -32.8
Maximum temperature: 3.9

Extreme low temperatures
 DATE
2010-01-01    -31.1
2010-01-02    -32.2
2010-01-08    -32.8
2010-01-09    -32.2
Name: TMIN, dtype: float64
```

Extreme high temperatures
 DATE
2010-01-14   3.9
2010-01-16   2.2
2010-01-17   3.3
2010-01-18   0.6
Name: TMAX, dtype: float64

Listing 6-4. The Function plot_temps That Produces the Plot As Shown in Figure 6-3

```
from matplotlib.ticker import MultipleLocator

def plot_temps(df, min_temp, max_temp, extreme_low_temps,
extreme_high_temps):
    ax1 = df.plot.line(y = ['TMAX', 'TMIN'],
            figsize = (12, 9),
            ylim = (1.3 * min_temp, 1.3 * max_temp),
            rot = 45, fontsize = 12, style = ['-r', '-b'],
            linewidth = 0.6,
            legend = False,
            x_compat = True)
    ax1.lines[0].set_label('Max. temperature')
    ax1.lines[-1].set_label('Min. temperature')
    ax1.set_title('Low and High Temperatures in January 2010\
nNorth Dakota, United States',
                fontsize = 20, y = 1.06)
    ax1.set_xlabel('Date', fontsize = 14, labelpad = 15)
    ax1.set_ylabel('Temperature [\u2103]', fontsize = 14)
    ax1.spines['right'].set_visible(False)
    ax1.spines['top'].set_visible(False)
    ax1.yaxis.set_minor_locator(MultipleLocator(5))
```

```
ax1.fill_between(df.index, df['TMAX'], df['TMIN'],
                 facecolor = 'lightgray', alpha = 0.25)

def fahrenheit_to_celisus(temp):
    return 1.8 * temp + 32

ax2 = ax1.twinx()
y_min, y_max = ax1.get_ylim()
ax2.set_ylim(fahrenheit_to_celisus(y_min), fahrenheit_to_
celisus(y_max))
ax2.set_ylabel('Temperature [\u2109]', fontsize = 14,
labelpad = 15)
ax2.spines['top'].set_visible(False)
ax2.yaxis.set_minor_locator(MultipleLocator(5))
for label in ax2.get_yticklabels():
    label.set_fontsize(12)

ax1.scatter(extreme_low_temps.index, extreme_low_temps,
            color = 'blue', marker = 'v', s = 100,
            label = 'Unusually low temperatures')
ax1.scatter(extreme_high_temps.index, extreme_high_temps,
            color = 'red', marker = '^', s = 100,
            label = 'Unusually high temperatures')
ax1.legend(loc = 4, frameon = False, title = 'Legend')
```

Observe the bold lines, which show how you switch from a higher Pandas visualization layer into the lower core matplotlib level. The raw interface is handy to fine-tune your diagram. It is also possible to go in the opposite direction by setting the parameter ax to your axis reference.

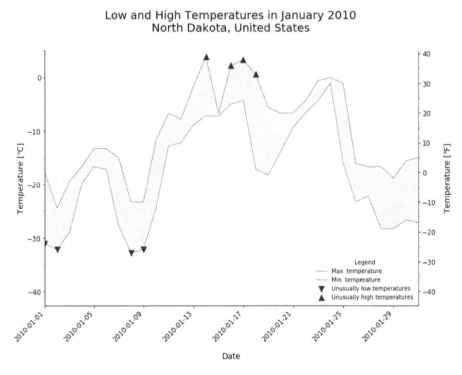

Figure 6-3. *The plot of low and high temperatures with markers for extreme temperatures (defined via thresholds)*

Closest Pair Case Study

Finding the closest pair is a well-known task from computational geometry with many applications. We have a set of points in the plane given by their x and y coordinates, represented as $P = \{(x,y) \mid x, y \in Z\}$. We need to find the closest pair of points $p_i \neq p_j$ whose Euclidian distance is defined as $\sqrt{(x_i - x_j)^2 + (y_i - y_j)^2}$. Let us assume that the number of points n is in the range of $[2, 10^7]$.

We will develop three core variants with different asymptotic behaviors: $O(n^2)$, $O(n\log^2 n)$, and $O(n \log n)$. You can refresh your memory about asymptotic notation and algorithms by consulting references [2]

218

through [4]. Afterward, we will bring down the constants to show how they affect the overall performance. Visualization will help us to illustrate the results more clearly. This is an example of using visualization to acquire deeper understanding of a program's run-time trait.

In the spirit of API-centric and object-oriented development, we will first create an abstract base class. In this manner, all children will simply inherit the same API, so clients would be able to smoothly change the implementation as desired (see also Exercise 6-1). Listing 6-5 shows this foundational class (look into the closest_pair package's base_closest_ pair.py file). Having a common API instigates experimentation, since you can switch implementations without disturbing the rest of your system. Moreover, a clear API is a prerequisite for successful code reuse.

Listing 6-5. Abstract Base Class Specifying the Shared API for All Subsequent Variants

```
"""The base class for implementing various variants to find the
closest pair."""

import abc
from typing import Tuple, Callable, TypeVar, Sequence, Generic
import numpy as np

Coordinates = TypeVar('Coordinates', Sequence[int], np.ndarray)

class BaseClosestPair(Generic[Coordinates], metaclass=abc.ABCMeta):
    """

    Finds the closest pair among 2D points given by their
    x and y coordinates. The distance is by default defined as
    a standard Euclidian distance.
```

219

```
    Args:
    x: the list of x coordinates of all points.
    y: the list of y coordinates of all points. The ordering of
        elements matches the list of x coordinates, i.e., the
        ith point is specified as (x[i], y[i]).
    """

    _x: Coordinates
    _y: Coordinates

    def __init__(self, x: Coordinates, y: Coordinates):
        assert len(x) >= 2 and len(x) == len(y)
        self._x = x
        self._y = y

    @property
    def x(self) -> Coordinates:
        """Gets the x coordinates of points."""
        return self._x

    @property
    def y(self) -> Coordinates:
        """Gets the y coordinates of points."""
        return self._y

    @staticmethod
    def load_from_stdin() -> Tuple[Coordinates, Coordinates]:
        """

        Loads points from standard input by enumerating x and y
        coordinates in succession.
        Each datum must be separated with space.
```

```
    Output:
    The tuple of x and y coordinates.
    """

    import sys

    data = sys.stdin.read()
    points = list(map(int, data.split()))
    x = points[1::2]
    y = points[2::2]
    return x, y

@staticmethod
def generate_points(n: int, seed: int) -> Tuple[Coordinates,
Coordinates]:
    """

    Generates random points for stress testing.

    Output:
    The tuple of x and y coordinates.

    Examples:
    >>> BaseClosestPair.gencrate_points(3, 10)
    ([227077737, -930024104, -78967768], [36293302,
    241441628, -968147565])
    """

    import random

    assert n >= 2
    random.seed(seed)
    x = [random.randint(-10**9, 10**9) for _ in range(n)]
    y = [random.randint(-10**9, 10**9) for _ in range(n)]

    return x, y
```

```
@staticmethod
def distance(x1: int, x2: int, y1: int, y2: int) -> float:
    """

    Returns the Euclidian distance between two points.

    Args:
    x1: the x coordinate of the first point.
    x1: the x coordinate of the second point.
    y1: the y coordinate of the first point.
    y2: the y coordinate of the second point.

    Output:
    The distance between points defined as the square root
    of the sum of squared differences of the matching
    coordinates.

    Examples:
    >>> BaseClosestPair.distance(1, 2, 1, 2)
    1.4142135623730951
    >>> BaseClosestPair.distance(1, 1, 1, 1)
    0.0
    """

    from math import sqrt

    return sqrt((x1 - x2)**2 + (y1 - y2)**2)

@abc.abstractmethod
def closest_pair(self, distance: Callable[[int, int, int,
int], float]) ↪
                -> Tuple[int, int, float]:
    """

    Returns back the tuple with indexes of closest points
    as well as their distance.
```

```
        Args:
            distance: the function that receives four parameters
            (x1, x2, y1, y2) and returns back the distance between
            these points.
            """

if __name__ == "__main__":
    import doctest
    doctest.testmod()
```

The closest_pair method can be easily customized by passing a new function to measure distances between points. The generate_points method can be used to stress test an implementation. To type check this module, issue the following command from Spyder's console (make sure you are in the closest_pair subfolder of this chapter's source code):

```
!mypy --ignore-missing-imports base_closest_pair.py
```

Our next task is to create a small test harness that will measure execution times and create a proper visualization. It turns out that there are available software tools for this purpose: cProfile (read more details at https://docs.python.org/3/library/profile.html) and SnakeViz (you may need to install it by executing conda install snakeviz from a command line). cProfile outputs data in a format that is directly consumable by SnakeViz.

Version 1.0

This version is a naive realization of the closest pair algorithm. It simply computes all pairwise distances and finds the minimum one. Listing 6-6 shows this version's source code.

Listing 6-6. The Brute-Force Implementation of This Problem

```python
"""Naive implementation of the closest pair algorithm."""

from typing import Tuple, Callable
from closest_pair.base_closest_pair import BaseClosestPair

class NaiveClosestPair(BaseClosestPair):
    def closest_pair(self, distance: Callable[[int, int, int,
    int], float] = ➥
                        BaseClosestPair.distance
                    ) -> Tuple[int, int, float]:
        """

        Iterates over all pairs and computes their distances.

        Examples:
        >>> x = [0, 3, 100]
        >>> y = [0, 4, 110]
        >>> ncp = NaiveClosestPair(x, y)
        >>> ncp.closest_pair()
        (0, 1, 5.0)
        """

        from math import inf

        n = len(self.x)
        min_distance = inf
        for i in range(n - 1):
            for j in range(i + 1, n):
                d = distance(self.x[i], self.x[j], self.y[i],
                self.y[j])
                if d < min_distance:
                    min_distance = d
                    p_i = i
                    p_j = j
```

```
        return p_i, p_j, min_distance
if __name__ == "__main__":
    import doctest
    doctest.testmod()
```

Tip Don't underestimate the value of such a simple but straightforward variant like `NaiveClosestPair`. It can be a very useful baseline both for performance comparisons and for assurance of correctness. Namely, it can be used together with the point generator to produce many low-dimensional test cases.

Let's profile one run using 100 autogenerated points. Execute the following statements inside Spyder's console (make sure that you are in this chapter's base source folder; see also Exercise 6-2):

```
>> from closest_pair import *
>> import cProfile
>> x, y = BaseClosestPair.generate_points(100, 1)
>> ncp = NaiveClosestPair(x, y)
>> %mkdir results
>> cProfile.run('ncp.closest_pair()', filename = 'results/
   ncp_100_stats')
```

We now should have the binary profile data inside the `results` subfolder. From a separate command-line shell, execute `snakeviz results/`. This starts up a local web server that listens by default on port 8080. The exact URL will be printed on a command line. Figure 6-4 shows the initial screen of SnakeViz.

SnakeViz

Search:

filename

ncp_100_stats

Showing 1 to 2 of 2 entries

SnakeViz Docs

Figure 6-4. *The entry page of SnakeViz, where you can select the* ncp_100_stats *file for visualization*

If you select the ncp_100_stats file in Figure 6-4, then you will see something like Figure 6-5 in the upper part of the page. I've hovered over the third rectangle from the top, which immediately displayed some more details about the execution of the closest_pair method. There are separate boxes for the overall signature (context) of a method and its constituent parts, including the body.

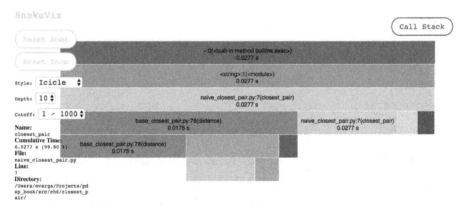

Figure 6-5. *The diagram is interactive*

Figure 6-6 shows the content of the lower part, which may look familiar to you (a classical tabular presentation of data). Having multiple views is very convenient, as you can smoothly switch between display formats to focus on the desired level of details. You may sort data by columns by clicking a column's label. Here, rows are ordered in descending order of `tottime`. The boxes are nested and organized into a hierarchy that reflects the call stack. Colors are additionally used to group elements. For example, the signature and body of the `closest_pair` method are colored pink. There are options on the right side to customize the view. Figure 6-7 shows the so-called *sunburst* diagram style.

Search:

ncalls	tottime ▾	percall	cumtime	percall	filename:lineno(function)
4950	0.009325	1.884e-06	0.01753	3.541e-06	base_closest_pair.py:78(distance)
1	0.007329	0.007329	0.02765	0.02765	naive_closest_pair.py:7(closest_pair)
4951	0.005063	1.023e-06	0.005063	1.023e-06	~:0(<built-in method builtins.hasattr>)
4951	0.001771	3.577e-07	0.006834	1.38e-06	<frozen importlib._bootstrap>:997(_handle_fromlist)
9901	0.001499	1.514e-07	0.001499	1.514e-07	base_closest_pair.py:28(x)
4950	0.001373	2.774e-07	0.001373	2.774e-07	~:0(<built-in method math.sqrt>)
9900	0.001293	1.306e-07	0.001293	1.306e-07	base_closest_pair.py:33(y)
1	2e-05	2e-05	0.02768	0.02768	~:0(<built-in method builtins.exec>)
1	9e-06	9e-06	0.02766	0.02766	<string>:1(<module>)
1	1e-06	1e-06	1e-06	1e-06	~:0(<built-in method builtins.len>)
1	1e-06	1e-06	1e-06	1e-06	~:0(<method 'disable' of '_lsprof.Profiler' objects>)

Showing 1 to 11 of 11 entries

Figure 6-6. *ncalls designates the number of calls made to a function (for example, 4950 = 100 × 99 / 2, which is the number of times* distance *was called), while* tottime *shows the total time spent inside a method*

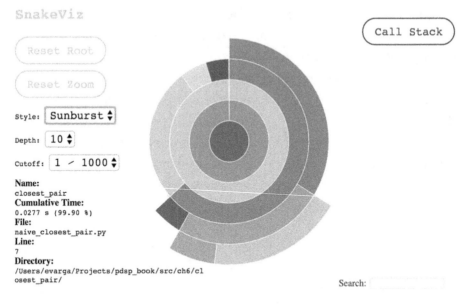

SnakeViz

Call Stack

Reset Root

Reset Zoom

Style: Sunburst ⬍

Depth: 10 ⬍

Cutoff: 1 ⁓ 1000 ⬍

Name:
closest_pair
Cumulative Time:
0.0277 s (99.90 %)
File:
naive_closest_pair.py
Line:
7
Directory:
/Users/evarga/Projects/pdsp_book/src/ch6/cl
osest_pair/

Search:

Figure 6-7. *Another way to visualize the calling hierarchy, which may be useful to see the big picture*

Let's try to run our quadratic algorithm on an input of size 3000. Here are the steps to execute:

```
>> x, y = BaseClosestPair.generate_points(3000, 1)
>> ncp = NaiveClosestPair(x, y)
>> cProfile.run('ncp.closest_pair()', filename = 'results/
   ncp_3000_stats')
```

This will take much more to finish. Navigate back to the folder page of SnakeViz by visiting http://127.0.0.1:<port number>/snakeviz/ results and selecting ncp_3000_stats. Sort the table by cumulative time in decreasing order and you will see that the running time is around 20 seconds (on my Mac it was 23.48s).[1] The previous run with 100 points took

---

[1]A box's size is commensurate to the cumulative time of the given method. This allows you to easily spot critical parts in your program. Don't forget that profiling adds some overhead.

only 0.0277s; a 30× increase of input size induced a roughly 900× upsurge of the running time. This is what $O(n^2)$ is all about. The exact times on your machine will be different, but the ratio will stay the same. Whatever you do to speed things up (for example, using Numba), you will remain at the same asymptotic level; in other words, the ratio of running times will mirror the quadratic dependence on input size. A pure Python code running a better algorithm will always outperform at one point an ultra-technologically boosted version that uses some slow algorithm.

Version 2.0

In this section we will lay out the $O(n\log^2 n)$ variant of the algorithm that uses the *divide-and-conquer* technique. This technique divides the initial problem into multiple subproblems of the same type and solves them independently. If any subproblem is still complex, then it is further divided using the same procedure. The chain ends when you hit a base case (i.e., a problem that is trivial to solve). Afterward, the results of subproblems are combined until the final answer is composed. A representative example of this methodology is the merge sort.

The major steps of this algorithm are as follows (see reference [2]):

1. Split the points into two halves based upon their x coordinate. More formally, if we denote the cutting coordinate as x, then we want to produce two sets of points, L (stands for left) and R (stands for right), such that all points of L have $x_i \leq x$ and all points of R have $x_i > x$.

2. Recursively treat groups L and R as separate subproblems and find their solutions as tuples (p_L, q_L, d_L) and (p_R, q_R, d_R), respectively. Take $d = \min(d_L, d_R)$ and remember from which group that minimum came.

3. Check if there is any point in L and R whose distance
    is smaller than d. Select all points of L and R whose
    x coordinate belongs to interval [x – d, x + d]. Sort
    these middle (M) points by their y coordinate (the
    ordering doesn't matter).

4. Iterate over this sorted list, and for each point,
    calculate its distance to at most seven subsequent
    points and maintain the currently found best pair.
    Let it be the triple (p_M, q_M, d_M).

5. The final solution is the minimum among pairs of
    L, M, and R. Recall that in step 2 we already found
    the minimum between L and R, so we just need one
    extra comparison with M.

At this level, an algorithm looks much like a recipe. You can even put
these steps into your source code as comments before doing any detailed
coding. As all steps are at a high level of detail, they nicely complement
low-level implementation stuff. This is the core idea behind pseudo-
code programming (planning with pseudo-code). Figure 6-8 visually
provides the proof of correctness of this algorithm. Proving a theorem
about your program's behavior is crucial when dealing with high-volume
data. Testing can reveal bugs in your code but cannot establish a ground
truth. Usually, it is hard to cover all possible scenarios purely with tests
(i.e., to find all equivalence classes in input). The downside of establishing
a mathematical proof (embodied inside formal software development
methods) is that it requires a specialized knowledge that most software
engineers lack. Moreover, you can only apply it to the core parts of your
system. For example, an operating system kernel is a feasible target for
such formal treatment.

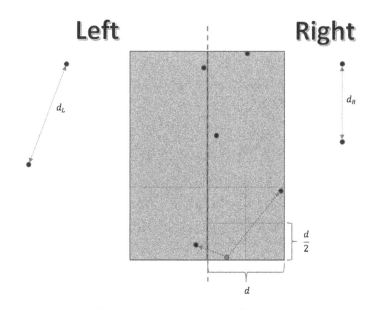

Figure 6-8. *Notice that inside any block of size d×d there can be at most four points*

Any of the OS kernel's subblocks of size $\dfrac{d}{2} \times \dfrac{d}{2}$ can only contain at most one point. In other words, if a square of dimensions d×d would contain five or more points, then some of its subblocks will surely host more than one point. Nonetheless, in this case we would reach a contradiction that d isn't correctly computed. Furthermore, any two points in this strip whose distance is < d must be part of the same rectangle of size 2d×d. Consequently, they definitely will be compared, which guarantees correctness. Recall that this rectangle may encompass at most eight points (including the reference point). Here, we will find a shorter distance than d, because the distance between the lowest-right point and subsequent left point is small.

Analysis of the Running Time

Assume that we are given n points in the plane. In step 1 we can find the median point that splits the set into two equally sized subsets, in O(n) time using a randomized divide-and-conquer algorithm. Creating subsets L and R is another O(n) time operation. After solving both subproblems of size $\dfrac{n}{2}$ the algorithm selects points belonging to the middle strip in O(n) time. Afterward, it sorts these filtered points in O(n log n) time. Finally, it computes distances inside this strip in O(n) time (for each point it does some fixed amount of work). Remember that for any function $f = kn$, $k \in \mathbb{N}$ it is true that $f = O(n)$; that is, it is bounded by some linear function. Therefore, irrespectively of how many times we do O(n) operations in succession or whether we do 3n, 5n, or 7n operations, their aggregate effect counts as O(n). All in all, the overall running time is defined by the recurrent formula $T(n) = 2T\left(\dfrac{n}{2}\right) + O(n\log n)$. We cannot directly apply the Master Theorem, although we may reuse its idea for analyzing our expression.

If we draw a recursion tree, then at some arbitrary level k, the tree's contribution would be $t_k = 2^k \dfrac{n}{2^k}\log\dfrac{n}{2^k} = 2^k\dfrac{n}{2^k}(\log n - k)$. The depth of the tree is log$n$, so by summing up all contributions from each level, we end up with $\displaystyle\sum_{k=0}^{\log n} t_k = n\log^2 n - n\sum_{k=0}^{\log n} k \le n\log^2 n - \dfrac{n}{2}\log^2 n = O(n\log^2 n)$.

In step 3 we repeatedly sort points by their y coordinate, and this amounts to that extra log$n$ multiplier. If we could eschew this sorting by only doing it once at the beginning, then we would lower the running time to O(n log n). This is evident from the new recurrence relation $T(n) = 2T\left(\dfrac{n}{2}\right) + O(n)$, since we may immediately apply the Master Theorem. Of course, we should be careful to preserve the y ordering of points during all sorts of splitting and filtering actions.

Version 3.0

To prevent disturbing the y ordering of points, we will sort them indirectly by finding the proper permutation of indices. All sorting functions actually compute this but usually don't explicitly return back. All access to original coordinates will happen through double indirection, as shown in Figure 6-9. Listing 6-7 presents the fast $O(n \log n)$ implementation of the previously described algorithm.

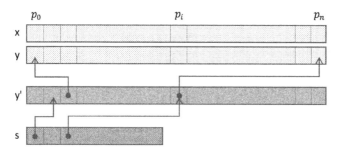

Figure 6-9. *The array* s *denotes the ordered indices of points belonging to the middle strip*

The array s comprises pointers into the sorted array of indices y', which finally point back into original positions. The complexity of accessing points in this manner will be hidden inside appropriate accessor methods.

Listing 6-7. Implementation of the FastClosestPair Class

```
"""Fast implementation of the closest pair algorithm."""

from typing import List, Tuple, Callable
from closest_pair.base_closest_pair import Coordinates,
BaseClosestPair

class FastClosestPair(BaseClosestPair):
    _y_prime: List[int]
```

```python
def _argsort_y(self) -> List[int]:
    """Finds the permutation of indices that arranges
    points by y coordinate."""

    return [t[0] for t in sorted(enumerate(self.y), key =
    lambda t: t[1])]

def _get_x(self, i: int, s: List[int]) -> int:
    return self.x[self._y_prime[s[i]]]

def _get_y(self, i: int, s: List[int]) -> int:
    return self.y[self._y_prime[s[i]]]

def __init__(self, x: Coordinates, y: Coordinates):
    super().__init__(x, y)
    self._y_prime = self._argsort_y()

def _selection(self, s: List[int], k: int) -> int:
    """Returns the x value of kth smallest point by x
    coordinate contained in s."""

    def split(v: int) -> Tuple[List[int], List[int],
    List[int]]:
        """Indirectly splits points in-place around value v
        into 2 sets (left and right)."""

        store = 0
        sl_idx = 0
        for i in range(len(s)):
            if self._get_x(i, s) < v:
                s[i], s[store] = s[store], s[i]
                store += 1
                sl_idx = store
        for i in range(store, len(s)):
            if self._get_x(i, s) == v:
```

```
            s[i], s[store] = s[store], s[i]
            store += 1
    return (s[:sl_idx], s[sl_idx:store], s[store:])

import random

v_idx = random.randrange(len(s))
v = self._get_x(v_idx, s)
sl, sv, sr = split(v)
sl_size = len(sl)
sv_size = len(sv)

if k <= sl_size:
    return self._selection(sl, k)
if k > sl_size and k <= sl_size + sv_size:
    return self._get_x(-1, sv)
return self._selection(sr, k - sl_size - sv_size)
```

```
@staticmethod
def _merge(sl: List[int], sr: List[int]) -> List[int]:
    """

    Merges the two sorted sublists into a new sorted list.
    The temporary storage may be allocated upfront as a
    further optimization.
    """

    sl_size = len(sl)
    sr_size = len(sr)
    s = [0] * (sl_size + sr_size)
    k = 0
    i = 0
    j = 0
```

```
        while i < sl_size and j < sr_size:
            if sl[i] <= sr[j]:
                s[k] = sl[i]
                k += 1
                i += 1
            else:
                s[k] = sr[j]
                k += 1
                j += 1
        while i < sl_size:
            s[k] = sl[i]
            k += 1
            i += 1
        while j < sr_size:
            s[k] = sr[j]
            k += 1
            j += 1

        return s

    def closest_pair(self,
                     distance: Callable[[int, int, int, int],
                     float] = BaseClosestPair.distance
                     ) -> Tuple[int, int, float]:
        """
```

Computes the minimum distance in O(n*log n) time.

Examples:
```
>>> x = [0, 3, 100]
>>> y = [0, 4, 110]
>>> fcp = FastClosestPair(x, y)
```

```
>>> fcp.closest_pair()
(0, 1, 5.0)
"""

from math import inf
```

def filter_points(s: List[int], d: float, x: int) -> List[int]:

```
    """Returns the list of point indexes that fall
    inside the [x-d, x+d] interval."""

    return [s[i] for i in range(len(s)) if abs(self._
    get_x(i, s) - x) <= d]
```

def find_nearest_neighbor(i: int, s: List[int]) -> Tuple[float, int, int]:

```
    """
    Finds the minimum distance between the current
    point i and next 7 seven subsequent points by y
    coordinate.
    """

    curr_x = self._get_x(i, s)
    curr_y = self._get_y(i, s)
    d = inf
    min_idx = i

    for j in range(i + 1, min(len(s), i + 7 + 1)):
        curr_d = distance(curr_x, self._get_x(j, s),
        curr_y, self._get_y(j, s))
        if curr_d < d:
            d = curr_d
            min_idx = j
    return d, s[i], s[min_idx]
```

```python
def find_minimum_distance(s: List[int]) -> Tuple[int,
int, float]:
    """Main driver function to find the closest pair."""

    if len(s) == 1:
        # We will treat the distance from a single
        point as infinite.
        return s[0], -1, inf
    if len(s) == 2:
        return s[0], s[1], distance(self._get_x(0, s),
                                    self._get_x(1, s),
                                    self._get_y(0, s),
                                    self._get_y(1, s))

    # This is the median value of input array x in
    regard of s.
    median_x = self._selection(s.copy(), len(s) // 2)

    # Separate points around median.
    sl = []
    sr = []
    for i in range(len(s)):
        if self._get_x(i, s) <= median_x:
            sl.append(s[i])
        else:
            sr.append(s[i])

    # Find minimum distances in left and right groups.
    p_l, q_l, d_l = find_minimum_distance(sl)
    p_r, q_r, d_r = find_minimum_distance(sr)
    if d_l < d_r:
        p_min, q_min = p_l, q_l
        d = d_l
```

```
            else:
                p_min, q_min = p_r, q_r
                d = d_r

            # Merge left and right indices keeping their sorted
            order.
            sm = FastClosestPair._merge(sl, sr)

            # Find the minimum distance inside the middle strip.
            sf = filter_points(sm, d, median_x)

            # Find the final minimum distance among three
            groups (left, middle, and right).
            d_m, p_m, q_m = min([find_nearest_neighbor(i, sf)
            for i in range(len(sf))])
            if d_m < d:
                return p_m, q_m, d_m
            else:
                return p_min, q_min, d

        p, q, d = find_minimum_distance(list(range(len
        (self._y_prime))))
        # We need to map back the point indices into their
        original base.
        return self._y_prime[p], self._y_prime[q], d

if __name__ == "__main__":
    import doctest
    doctest.testmod()
```

Obviously, this version is much more complex than our naive attempt in version 1.0. Let's see how it performs on the same dataset of 3000 points. The steps to create a new report are as follows (I assume that you are continuing the last console session, so you already have arrays x and y defined):

```
>> from closest_pair import FastClosestPair
>> fcp = FastClosestPair(x, y)
>> cProfile.run('fcp.closest_pair()', filename = 'results/
   fcp_3000_stats')
```

Figure 6-10 shows the diagram of this run. If you scroll down in SnakeViz, you will see that the cumulative time has dropped well below one second. Once you have a good algorithm, then you can be sure that your solution is scalable. Of course, even this improvement is far from the absolute lower bound, after you reduce the constant factor and apply parallel processing (see also Exercise 6-3). Nevertheless, such additional tweaks are meaningful only after you have already obtained a good asymptotic behavior.

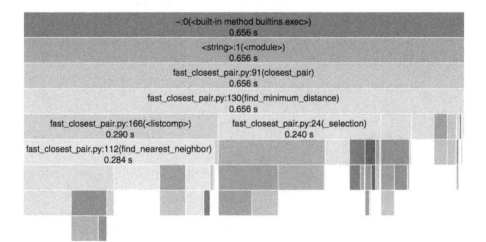

Figure 6-10. *The picture only shows the Icicle-style diagram. Apparently, there are many additional details, since the code base is more involved.*

Enquiry of Algorithms Evolution

The differences between the brute-force and sophisticated versions are indeed striking from multiple viewpoints. Besides exponential speedup (from n × n to n × $log\ n$), there is a proportional increase in complexity. Nonetheless, by focusing on API-centric development, clients of our classes are totally shielded from inner details. Both versions can be used in exactly the same fashion by invoking public methods specified in the base class. This is once again a testimony that sound software engineering skills are real saviors in data science endeavors.

The moral of the story is that purely technological hocus-pocus cannot get you far away. Sure, we demonstrated Numba in the previous chapter, but real breakthroughs can only happen by heralding mathematics. This is generally true for computer science, not only data science, although in case of the latter, the effect is even more amplified, as no naive approach can cope with Big Data.

Besides time-related performance optimization, you may want to measure memory consumption and other resources. For this you also have lots of offerings. Take a look at the `memory-profiler` package (see `https://pypi.org/project/memory-profiler`). There are also suitable magic command wrappers `%memit` and `%mprun`. You can use them by first issuing `%load_ext memory_profiler`. The same is true for time profiling using the `line_profiler` module (see `https://pypi.org/project/line_profiler`) with the `%lprun` magic command.

Interactive Information Radiators

Most often we produce active reports that summarize our findings regarding some static dataset. A common platform to deliver live reports of this kind is the standard Jupyter notebook. However, in some situations the supporting dataset is continually changing over time. Forcing users to rerun notebooks and pick up the latest dataset isn't a scalable solution (let alone resending them a refreshed notebook each time). A better option is to set up a pipeline and let dynamic reports be created automatically. This is the idea behind information radiators, which are also known as *dashboards*.

We can differentiate two major categories of fluctuating data: batch-oriented data, which is accumulated and processed at regular intervals (for example, hourly, daily, weekly, etc.), and streaming data, which is updated continuously. The concomitant dashboard must reflect this rate of change and expose solely crucial business metrics for monitoring and fast decision making. Any superfluous detail will just deter users' attention. Of course, which type of dashboard you should build also depends on your pipeline's capabilities.

Suppose that you collect electrical power production data from various power plants, including those using wind and solar energy. The following information may be useful to efficiently gauge the current and near future state of the power network:

- Current energy production and consumption

- Energy production by different plants (nuclear, thermal, water, wind, solar, etc.)

- The short-term weather forecast, which is an outside factor that could considerably impact the efficacy of wind and solar plants

Once you decide what to put onto your dashboard, then you must contemplate how to convey things visually in the best possible fashion. Domain knowledge is very important here. It turns out that a Sankey diagram is a perfect match. It visualizes the flow between nodes in an acyclic directed graph (a power network may be represented in such form). Figure 6-11 shows an example of this graph.

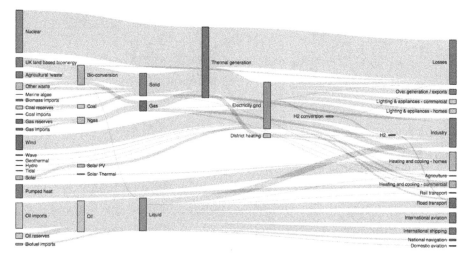

Figure 6-11. *A contrived scenario of energy production and consumption in the UK in the future (see* https://github.com/d3/ d3-sankey*)*

The Sankey diagram is by default interactive, and you can select more details about particular branches by clicking or hovering over it. Now imagine this diagram is part of your widely accessible notebook that is regularly updated by your pipeline. This is the core idea behind dashboarding with notebooks (see reference [5]).

The key ingredients for realizing the vision of interactive information radiators (let's assume Jupyter Notebook as our document format) are as follows:

- Regularly updated and published dataset(s) (for trying out things, you may use some datasets offered by Kaggle; see reference [5])

- A dashboarding pipeline that would pick up relevant datasets and rerun your notebook

- An interactive notebook, most likely utilizing Altair (see
  `https://altair-viz.github.io`), Plotly (see `https://plot.ly`), Bokeh (see `https://bokeh.pydata.org`),
  or a similar framework, as you probably don't want to
  code stuff from scratch

The Power of Domain-Specific Languages

A domain-specific language (DSL) aids productivity and quality by
relieving you from mundane implementation details, so that you can
concentrate on business-related elements only. Visualization is a well-
bounded domain and is amenable to being supported by a DSL. Altair is
a framework that provides such a DSL and enables you to do declarative
visualization in Python. This capability is especially vital for crafting
complex interactive workflows in the form of dashboards. Spending time
on low-level minutiae is unproductive and is a common source of error.
Let me demonstrate the power of DSL by introducing Anscombe's quartet.
Besides its technical aspects, this example is also instructive to emphasize
the prominence of visualization. Listing 6-8 shows how you might address
this task using only matplotlib. It still looks like ordinary Python code
despite using matplotlib's object-oriented API.

Listing 6-8. Without Any High-level Framework, There Are Lots of
Moving Targets to Keep Under Control

```
import numpy as np
import matplotlib.pyplot as plt

quartets = np.asarray([
    (
    [10.0, 8.0, 13.0, 9.0, 11.0, 14.0, 6.0, 4.0, 12.0, 7.0, 5.0],
    [8.04, 6.95, 7.58, 8.81, 8.33, 9.96, 7.24, 4.26, 10.84,
    4.82, 5.68]
    ),
```

```
    (
      [10.0, 8.0, 13.0, 9.0, 11.0, 14.0, 6.0, 4.0, 12.0, 7.0, 5.0],
      [9.14, 8.14, 8.74, 8.77, 9.26, 8.10, 6.13, 3.10, 9.13,
      7.26, 4.74]
    ),
    (
      [10.0, 8.0, 13.0, 9.0, 11.0, 14.0, 6.0, 4.0, 12.0, 7.0, 5.0],
      [7.46, 6.77, 12.74, 7.11, 7.81, 8.84, 6.08, 5.39, 8.15,
      6.42, 5.73]
    ),
    (
      [8.0, 8.0, 8.0, 8.0, 8.0, 8.0, 8.0, 19.0, 8.0, 8.0, 8.0],
      [6.58, 5.76, 7.71, 8.84, 8.47, 7.04, 5.25, 12.50, 5.56,
      7.91, 6.89]
    )
])

roman = ['I', 'II', 'III', 'IV']

fig = plt.figure(figsize = (12, 9))
fig.suptitle("Anscombe's Quartets", fontsize=16)
axes = fig.subplots(2, 2, sharex = True, sharey = True)

for quartet in range(quartets.shape[0]):
    x, y = quartets[quartet]
    coef = np.polyfit(x, y, 1)
    reg_line = np.poly1d(coef)

    ax = axes[quartet // 2, quartet % 2]
    ax.plot(x, y, 'ro', x, reg_line(x), '--k')
    ax.set_title(roman[quartet])
    ax.set_xlim(3, 19.5)
    ax.set_ylim(2, 13)
```

```
    # Print summary statistics for the current dataset
    print("Quartet:", roman[quartet])
    print("Mean X:", x.mean())
    print("Variance X:", x.var())
    print("Mean Y:", round(y.mean(), 2))
    print("Variance Y:", round(y.var(), 2))
    print("Pearson's correlation coef.:", round(np.corrcoef(x,
    y)[0][1], 2))
    print()

plt.show()
```

The program will print the following statistics about each quartet (observe that they are all the same):

```
Quartet: I
Mean X: 9.0
Variance X: 10.0
Mean Y: 7.5
Variance Y: 3.75
Pearson's correlation coef.: 0.82

Quartet: II
Mean X: 9.0
Variance X: 10.0
Mean Y: 7.5
Variance Y: 3.75
Pearson's correlation coef.: 0.82

Quartet: III
Mean X: 9.0
Variance X: 10.0
Mean Y: 7.5
Variance Y: 3.75
Pearson's correlation coef.: 0.82
```

```
Quartet: IV
Mean X: 9.0
Variance X: 10.0
Mean Y: 7.5
Variance Y: 3.75
Pearson's correlation coef.: 0.82
```

Nonetheless, the differences are obvious once you visualize these datasets, as shown in Figure 6-12. Listing 6-9 shows the snippet that achieves a similar visualization, but this time using Altair's DSL (see also reference [6] about DSLs). The code is compact without unnecessary low-level details. With a DSL, you only specify what you want to accomplish and let the engine decide the best tactic to deliver the requested output. You need to run this code from a JupyterLab notebook by simply pasting it into a code cell.

Figure 6-13 shows its graphical output (see also Exercise 6-4).

Listing 6-9. Visualization Using Altair's DSL

```python
import altair as alt
from vega_datasets import data

source = data.anscombe()

base = alt.Chart(
    source, title = "Anscombe's Quartets"
).mark_circle(color = 'red').encode(
    alt.X('X', scale = alt.Scale(zero = True)),
    alt.Y('Y', scale = alt.Scale(zero = True)),
    column = 'Series'
).properties(
    width = 150,
    height = 150
).interactive()

base
```

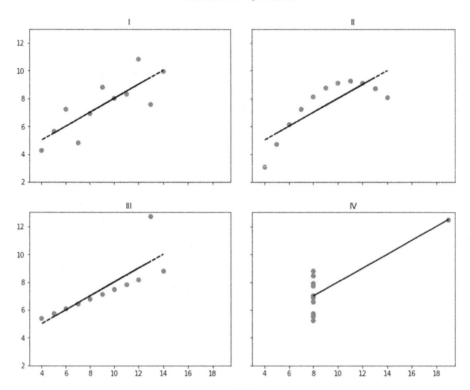

Figure 6-12. *The four quartets are definitely different, although they share the same summary statistics*

You will need to install Altair via the conda package manager by issuing the following statement (it will also install the useful vega_datasets, which already contains data for Anscombe's quartet):

```
conda install -c conda-forge altair vega_datasets
```

Tip   You may want to create a snapshot of the dashboard's state at some moment. All charts produced by Altair include a drop-down menu to save diagrams as images or export them into Vega format.

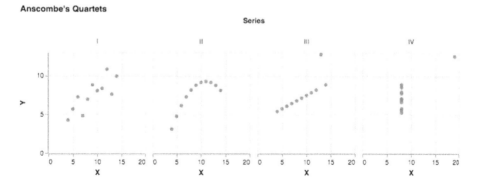

***Figure 6-13.** Anscombe's quartet using default arrangement in Altair*

The plot is interactive, and you may pan and zoom each diagram. In interactive dashboards you may want to minimize textual information as well as hide code cells in your notebook. These steps bring your graphical elements into the foreground.

Another advantage of using declarative frameworks is that your notebook will be devoid of device-specific stuff. The whole point of having dashboards is to allow people to access data anytime and everywhere. This entails that many people will use portable devices to browse data. Your dashboard must fit the screen perfectly regardless of whether the user is viewing it on a computer, laptop, tablet, or smart phone. Taking care of all this should be left to the framework.

EXERCISE 6-1. CREATE A FACTORY

Clients shouldn't instantiate classes directly; ideally, they shouldn't even be aware of concrete implementation classes. This is the doctrine behind the principle to program against interfaces instead of implementation artifacts. There are two design patterns associated with realizing this vision: Factory Method and Abstract Factory. The latter makes another step in this direction to isolate clients even from factory implementations.

Refactor the code base and introduce these factory entities. In this manner, clients would simply pass the relevant portfolio to a factory (describing the properties of the desired implementation) and receive an object exposed through the shared API (interface).

EXERCISE 6-2. IMPLEMENT A TEST HARNESS

Instead of trying out different versions from the console, which is a bit tedious, create a new utility class to test the system. You should encompass the following steps in your flow:

1. Get a fresh instance of points by calling the `BaseClosestPair.generate_points` method (the number of points should be capped at 1000).

2. Find the correct solution using some reliable oracle (for example, by using the trivial `NaiveClosestPair` class).

3. Call the target implementation with the same dataset.

4. Compare and report the result.

Thanks to the existence of the common API, your test harness may accept any target implementation of this type. This will create an opportunity for future optimization attempts. Try to produce a `ParallelClosestPair` class. Observe that our algorithm is inherently parallelizable, since left and right groups may be processed independently.

EXERCISE 6-3. PRODUCE AN ULTRA-FAST VARIANT

Thus far, we have used only pure Python constructs. Unfortunately, this does not usually result in a top-performing solution. Lists are very slow, since they are flexible and heterogeneous. Using the built-in array would be a better option for our case. Nonetheless, even this is not enough. Real speedup would happen by switching to Numpy arrays and compiled code.

Create an `UltraFastClosestPair` class that utilizes Numpy arrays; I recommend that you do Exercise 6-2 before embarking on this exercise. Notice that some stuff will automatically disappear from the current source. For example, Numpy provides the `argsort` method out of the box. Furthermore, you can use libraries around Numpy. The `sortednp` package (see https://pypi.org/project/sortednp) is very handy, as it implements the `merge` method in O(n) time.

EXERCISE 6-4. PRACTICE OVERLAYING CHARTS IN ALTAIR

In our matplotlib-based code, notice how the regression line is combined with the base scatter plot; see the bold `plot` instruction in Listing 6-4. Everything was bundled together without having the notion of a chart as a separate abstraction. This is completely different in Altair, where charts behave like self-contained entities. If you have handles to multiple charts, then you can mix them in an ad hoc manner utilizing the overloaded + operator.

Add regression lines to the chart referenced from variable `base`. You would still use Numpy to compute the coefficients. Take a look at the following tutorial for more details: https://altair-viz.github.io/gallery/poly_fit.html. Furthermore, visit https://stackoverflow.com/questions/50164001/multiple-column-row-facet-wrap-in-altair for help on how to simulate subplots from matplotlib.

Summary

It is impossible to iterate over all kinds of visualizations, since there are myriads of them; you may want to visit the D3 visualization gallery at `https://github.com/d3/d3/wiki/Gallery`, which collects many types of visualizations in a single place. The most important thing to remember is to always take into account the goals of stakeholders and contemplate how graphical presentation of information may help them. In other words, everything you do must have a good reason and be part of the context. Don't make fancy displays just for the sake of entertainment.

Interactive displays are very powerful, as they combine analytics with graphical presentation. When the underlying data is changing rapidly, users typically like to monitor those changes and perhaps perform actions pertaining to what-if scenarios. Therefore, center your interaction around those frequent activities, so that users may quickly play around with data in a focused manner. Making powerful interactive dashboards demands knowledge and proficiency about user experience design, which is quite extensive territory. Again, we have revisited the same conclusion that any nontrivial data science project is a team effort.

Dashboarding requires automation, and cloud-based services are well suited for this purpose. They provide the necessary scalable and fault-tolerant infrastructure so that you can set up your pipeline smoothly. You might want to look into two prominent options: PythonAnywhere (see `https://www.pythonanywhere.com`) and Google Cloud Platform (see `https://cloud.google.com`). The former relies on Amazon Web Services (AWS). Of course, you can also create an on-premises solution using OpenStack (see `https://www.openstack.org`). You can quickly get a jump-start on dashboarding by using Kaggle's API and guidelines from reference [5].

References

1. Michael Dubakov, "Visual Encoding," `https://www.targetprocess.com/articles/visual-encoding`, Sept. 2012.

2. Sanjoy Dasgupta, Christos Papadimitriou, and Umesh Vazirani, *Algorithms*, McGraw-Hill Education, 2006.

3. Thomas H. Cormen, Charles E. Leiserson, Ronald L. Rivest, and Clifford Stein, *Introduction to Algorithms, Third Edition*, MIT Press, 2009.

4. Tom Cormen and Devin Balkcom, "Algorithms," Khan Academy, `https://www.khanacademy.org/computing/computer-science/algorithms`.

5. Rachael Tatman, "Dashboarding with Notebooks: Day 1," `https://www.kaggle.com/rtatman/dashboarding-with-notebooks-day-1/notebook`, 2019.

6. Martin Fowler, *Domain-Specific Languages*, Addison-Wesley Professional, 2010.

CHAPTER 7

Machine Learning

Machine learning is regarded as a subfield of artificial intelligence that deals with algorithms and technologies to squeeze out knowledge from data. Its fundamental ingredient is Big Data, since without help of a machine, our attempt to manually process huge volumes of data would be hopeless. As a product of computer science, machine learning tries to approach problems algorithmically rather than purely via mathematics. An external spectator of a machine learning module would admire it as some sort of magic happening inside a box. Eager reductionism may lead us to say that it all is just "bare" code executed on a classical computer system. Of course, such a statement would be an abomination. Machine learning does belong to a separate branch of software, which learns from data instead of blindly following predefined rules. Nonetheless, for its efficient application, we must know how and what such algorithms learn as well as what type of algorithm(s) to apply in a given context. No machine learning system can notice that it is misappropriated. The goal of this chapter is to lay down the foundational concepts and principles of machine learning exclusively through examples.

© Ervin Varga 2019
E. Varga, *Practical Data Science with Python 3*,
https://doi.org/10.1007/978-1-4842-4859-1_7

There are multiple ways to group machine learning algorithms. We can differentiate between the following three learning styles:

> **Supervised learning**: Here an algorithm is exercised on known observations until it achieves a desirable level of performance. The main challenge is to acquire enough high-quality marked data for appropriate training. Some members of this group are linear regression, logistic regression, support vector machine, naive Bayes classifier, etc.

> **Unsupervised learning**: These algorithms try to autonomously discover hidden structures in data for the purpose of grouping them, finding interesting relationships (a.k.a. association rule learning), or reducing inherent dimensionality (describe effects with fewer features). Some members of this group are K-Means clustering, principal component analysis, manifold learning, Aprori algorithm, etc.

> **Semi-supervised learning**: These algorithms try to decipher hidden structures based on guidance from labeled specimens. It isn't uncommon that unsupervised learning algorithms are run on half-marked data, as a preprocessing step, to label remaining data points (a technique known as *label propagation*).

Besides grouping algorithms, we can also discern various learning methods as follows:

> **Full-batch learning (a.k.a. statistical learning)**: We feed an algorithm all training data at once. After initial training, model parameters remain fixed. This scheme is also most popular for demonstrating various algorithms in action due to its simplicity.

Mini-batch learning: We feed an algorithm in a chunked manner. For example, with an enormous training sample, the standard gradient descent optimization method is prohibitive. This is where the stochastic gradient descent alternative becomes attractive, which works with a reduced chunk at any moment of time.

Online learning: This is an extreme case of mini-batching in which the batch size is reduced to a single observation. The usual setup is that the system is warmed up on historical data and left to update its parameters while running in production. This learning method has many subvariants, which are listed below as separate groups to avoid nesting.

Streaming: Here, an algorithm also works on a single observation at a time but cannot revisit past records (generic online learning algorithms are allowed to go over data multiple times). It is OK for a streaming system to cache recent items or keep running statistics (like moving average), although these are miniscule in comparison to the training corpus used in previous sets. Finally, many streaming approaches actually rely on micro-batching (with tiny batches). They are still called streaming since they don't pass multiple times over older batches.

Active learning: This method actively seeks feedback from other entities while running. For example, an online learning–based spam filter may incorporate user actions (declassify a piec of spam as normal, or vice versa) to continuously update itself. A semi-supervised system could ask for help to label conflicting (confusing) observations.

Reinforcement learning: This special dynamic learning method builds upon interaction between the system and its environment. Using an efficient feedback loop and rewarding mechanism (with positive and negative rewards), an algorithm learns what actions are proper in a given context. Robots typically learn in this manner.

We can also embark on separating algorithms by their similarity (e.g., neural networks, spectral methods, Bayesian networks, etc.). Nevertheless, any classification can only broadly enumerate knowledge areas, with blurred borders and overlaps. However, some notions are shared by all of them. These will be the topic in the rest of this chapter.

Irrespectively of what approach you choose, at one point in time, you will need to peek under the hood of many algorithms to tweak advanced parameters, combine them to work in a unified fashion (a.k.a. ensemble), or connect them as a pipeline for staged processing. You cannot eventually escape deep mathematics. The good news is that you can survive this encounter in incremental fashion.

Exposition of Core Concepts and Techniques

This section gradually introduces common machine learning concepts using *ordinary least squares regression*, which is simple to comprehend yet powerful enough to serve as a basis for discussion. It establishes a linear relationship between features (predictors) to predict an output value (response). The features themselves may be of arbitrary degree and complexity (for example, they could be polynomial terms). Listing 7-1 shows the data_generator.py module, which produces features and outputs based on various criteria by simulating fake "real world" processes. Listing 7-1 provides the scaffolding for our demo harness to exhibit the following notions: real world, training process, process parameters, runtime model, runtime parameters, estimators, evaluation metrics (loss function, mean squared error, explained variance), overfitting, underfitting, feature interaction, collinearity, and regularization. My intention here is to err on the side of oversimplification to convey essential ideas. There are a lot of excellent books about machine learning with a gamut of complex mathematics behind the scenes; the field is immense.

Listing 7-1. data_generator.py Module

```python
import numpy as np
import pandas as pd

def generate_base_features(sample_size):
    x_normal = np.random.normal(6, 9, sample_size)
    x_uniform = np.random.uniform(0, 1, sample_size)
    x_interacting = x_normal * x_uniform
    x_combined = 3.6 * x_normal + np.random.exponential(2/3,
    sample_size)
    x_collinear = 5.6 * x_combined

    features = {
        'x_normal': x_normal,
        'x_uniform': x_uniform,
        'x_interacting': x_interacting,
```

```
        'x_combined': x_combined,
        'x_collinear': x_collinear
    }
    return pd.DataFrame.from_dict(features)

def identity(x):
    return x

def generate_response(X, error_spread, beta, f=identity):
    error = np.random.normal(0, error_spread, (X.shape[0], 1))
    intercept = beta[0]
    coef = np.array(beta[1:]).reshape(X.shape[1], 1)
    return f(intercept + np.dot(X, coef)) + error
```

Our intimate knowledge about each underlying data generator process allows us to illuminate concepts in an exact manner. The real world simulated by generate_response has the following vector form: $y = f(\beta_0 + X\beta_{1:n}) + \varepsilon$ (see also the sidebar "Linear Regression Varieties"), where f is a discretionary function. The error term represents an inherent *noise*, which isn't encompassed by the model. We cast it as a Gaussian random variable $\varepsilon \sim \mathcal{N}(\mu = 0, \sigma = error\_spread)$; that is, a term that symmetrically fluctuates around the output following the Normal distribution. For simplicity, we produce *homoscedastic* outputs, meaning the errors are uncorrelated and uniform (they don't depend on X). The vector β is arbitrary.

An external observer can only see records (x_i, y_i) from this world, where $i \in [1, sample\ size]$. The art, engineering, and science is to reconstruct real-world phenomena by using only one or more samples, as shown in Figure 7-1. We proceed by *assuming* a specific model (for example, linear with a given error distribution). Afterward, we try to figure out model parameters (like vector β and σ) and seek proper features (this obviously has an impact on parameters). All in all, we have lots of stuff to presume and calculate.

Figure 7-1. *[1]From outside we can only gather observations and try to establish sound relationships between features and outputs*

Establishing relationships between features and output allows us to either predict future outcomes or better understand what is going on in the system. In Figure 7-1, the frontal digits represent assumed predictors and fitted values (what we closely examine and crunch in production), while the digits on the globe represent real-world data. We can state that our model closely resembles the real world if there is an acceptable match between them.

---

[1]You can find many attractive and free pictures on Pixabay (visit `https://pixabay.com`), like this digital world.

LINEAR REGRESSION VARIETIES

The generic form of a linear regression is given by $y = m(x) + \varepsilon$. The conditional expectation function m is E[y|x], if E[ε|x]=0 (when the homoscedasticity property holds). Depending on m's structure, we can discern the following types of linear regressions:

- **Parametric**: $m = \beta_0 + X\beta_{1:n}$ is smooth, defined, and easily interpretable.

- **Nonparametric**: m is smooth, must be discovered from data, flexible, and hard to interpret.

- **Semiparametric**: Somewhere between the preceding two cases (partially structured).

You must evaluate the cost-benefit ratio before deciding which version to use. For example, if you have outliers (extreme but valid observations that you cannot eradicate), then simple parametric linear regression isn't a good choice. The nonparametric version is more robust and less vulnerable regarding outliers.

Imagine that you encounter a helpful wizard, who reveals everything about the real world's processes. With that knowledge, how accurately may you predict future outcomes? You definitely cannot talk in absolute terms, as you have to contend with randomness in making your predictions; you may only speak about probabilities of seeing particular responses. In other words, you can develop a conditional probability distribution of an output given an observation. This can be expressed as $P(y|X) = \mathcal{N}\left(f\left(\beta_0 + \beta_{1:n}X\right), \sigma\right)$.

The truth is, you are not straying far from reality (and sanity) if you are hoping for such a wizard. The scikit-learn framework (https://scikit-learn.org) is a wizard in its own right. You just need to carefully interpret scikit-learn's words and never misapply the power bestowed on you. The aim of this section is to shed some light on necessary communication skills and language so that you become this wizard's beloved apprentice.

Listing 7-2 shows the `scikit-learn` package in action with some visualization of what happens under the hood. The `observer.py` module contains functions to recover parameters and demonstrate various effects pertaining to training, testing, and evaluation. You don't need to fully understand the auxiliary display code (see the `explain_sse` and `plot_mse` functions) to follow along. The `train_model` function recovers the real world's parameters solely by using observations (a.k.a. labeled training data). The `evaluate_model` function checks the validity of the model by calculating the *mean squared error (MSE)* and *explained variance score* using test data. In practice, you should always select performance criterion to align with objectives of the problem. Out-of-context measurements are distorting.

Listing 7-2. `observer.py` Module

```python
import numpy as np
import pandas as pd
import seaborn as sns
import matplotlib.pyplot as plt

plt.style.use('seaborn-whitegrid')

def train_model(model, X_train, y_train):
    model.fit(X_train, y_train)

def evaluate_model(model, X_test, y_test, plot_residuals=False,
title=""):
    from sklearn.metrics import mean_squared_error, explained_
    variance_score

    y_pred = model.predict(X_test)
```

```
    if plot_residuals:
        _, ax = plt.subplots(figsize=(9, 9))
        ax.set_title('Residuals Plot - ' + title, fontsize=19)
        ax.set_xlabel('Predicted values', fontsize=15)
        ax.set_ylabel('Residuals', fontsize=15)
        sns.residplot(y_pred.squeeze(), y_test.squeeze(),
                      lowess=True,
                      ax=ax,
                      scatter_kws={'alpha': 0.3},
                      line_kws={'color': 'black', 'lw': 2,
                      'ls': '--'})

    metrics = {
        'explained_variance': explained_variance_score(y_test,
        y_pred),
        'mse': mean_squared_error(y_test, y_pred)
    }
    return metrics

def make_poly_pipeline(model, degree):
    from sklearn.pipeline import make_pipeline
    from sklearn.preprocessing import PolynomialFeatures

    return make_pipeline(PolynomialFeatures(degree=degree,
    include_bias=False), model)

def print_parameters(linear_model, metrics):
    print('Intercept: %.3f' % linear_model.intercept_)
    print('Coefficients: \n', linear_model.coef_)
    print('Explained variance score: %.3f' %
    metrics['explained_variance'])
    print("Mean squared error: %.3f" % metrics['mse'])
```

```python
def plot_mse(model, X, y, title, error_spread):
    def collect_mse():
        from sklearn.model_selection import train_test_split
        from sklearn.model_selection import cross_val_score

        metrics_all = []
        for train_size_pct in range(10, 110, 10):
            X_train, X_test, y_train, y_test = \
                train_test_split(X, y, shuffle=False, train_
                size=train_size_pct / 100)
            metrics_current = dict()
            metrics_current['percent_train'] = train_size_pct
            train_model(model, X_train, y_train)
            metrics_train = evaluate_model(model, X_train, y_
            train)
            metrics_current['Training score'] = metrics_
            train['mse']
            metrics_cv = cross_val_score(
                model,
                X_train, y_train,
                scoring='neg_mean_squared_error', cv=10)
            metrics_current['CV score'] = -metrics_cv.mean()
            if X_test.shape[0] > 0:
                metrics_test = evaluate_model(model, X_test,
                y_test)
                metrics_current['Testing score'] = metrics_
                test['mse']
            else:
                metrics_current['Testing score'] = np.NaN
            metrics_all.append(metrics_current)
        return pd.DataFrame.from_records(metrics_all)
```

```
import matplotlib.ticker as mtick

df = collect_mse()
error_variance = error_spread**2
ax = df.plot(
    x='percent_train',
    title=title,
    kind='line',
    xticks=range(10, 110, 10),
    sort_columns=True,
    style=['b+--', 'ro-', 'gx:'],
    markersize=10.0,
    grid=False,
    figsize=(8, 6),
    lw=2)
ax.set_xlabel('Training set size', fontsize=15)
ax.xaxis.set_major_formatter(mtick.PercentFormatter())
y_min, y_max = ax.get_ylim()
# FIX ME: See Exercise 2!
ax.set_ylim(max(0, y_min), min(2 * error_variance, y_max))
ax.set_ylabel('MSE', fontsize=15)
ax.title.set_size(19)

# Draw and annotate the minimum MSE.
ax.axhline(error_variance, color='g', ls='--', lw=1)
ax.annotate(
    'Inherent error level',
    xy=(15, error_variance),
    textcoords='offset pixels',
    xytext=(10, 80),
    arrowprops=dict(facecolor='black', width=1, shrink=0.05))
```

```python
def explain_sse(slope, intercept, x, y):
    # Configure the diagram.
    _, ax = plt.subplots(figsize=(7, 9))
    ax.set_xlabel('x', fontsize=15)
    ax.set_ylabel('y', fontsize=15)
    ax.set_title(r'$SSE = \sum_{i=1}^n (y_i - \hat{y}_i)^2$',
    fontsize=19)
    ax.grid(False)
    ax.spines["top"].set_visible(False)
    ax.spines["right"].set_visible(False)
    ax.tick_params(direction='out', length=6, width=2,
    colors='black')

    # Show x-y pairs.
    ax.scatter(x, y, alpha=0.5, marker='x')

    # Draw the regression line.
    xlims = np.array([np.min(x), np.max(x)])
    ax.plot(xlims, slope * xlims + intercept, lw=2, color='b')

    # Draw the error terms.
    for x_i, y_i in zip(x, y):
        ax.plot([x_i, x_i], [y_i, slope * x_i + intercept],
            color='r', lw=2, ls='--')
```

Notice the advantage of having a unified API (like, fit, predict, etc.) to work with various models. Even pipelines are handled in the same manner. Listing 7-3 shows the session.py module's demo_metrics_and_mse function. It depicts steps to reconstruct parameters from observations using various noise levels. In the absence of noise we have a deterministic linear regression. As noise increases, the estimation of parameters deteriorates.

Listing 7-3. First Part of session.py Module (Some Imports Are for Later Use)

```python
import warnings
warnings.simplefilter(action='ignore', category=FutureWarning)

import numpy as np
import pandas as pd
from sklearn.linear_model import LinearRegression

from data_generator import *
from observer import *

def set_session_seed(seed):
    np.random.seed(seed)      # Enables perfect reproduction of
                              # published results.

def demo_metrics_and_mse():
    set_session_seed(100)
    X = generate_base_features(1000)[['x_normal']]
    for noise_level in [0, 2, 15]:
        y = generate_response(X, noise_level, [-1.5, 4.1])
        model = LinearRegression()
        train_model(model, X, y)
        metrics = evaluate_model(model, X, y)

        print('\nIteration with noise level: %d' % noise_level)
        print_parameters(model, metrics)

        # Visualize the regression line and error terms.
        if noise_level == 15:
            slope = model.coef_[0][0]
            intercept = model.intercept_
            explain_sse(slope, intercept, X[:15].values, y[:15])
```

The warning module is used to silences FutureWarning warning types, thus removing clutter from output. The following is the printout from executing demo_metrics_and_mse():

```
Iteration with noise level: 0
Intercept: -1.500
Coefficients:
 [[4.1]]
Explained variance score: 1.000
Mean squared error: 0.000

Iteration with noise level: 2
Intercept: -1.470
Coefficients:
 [[4.0925705]]
Explained variance score: 0.997
Mean squared error: 4.187

Iteration with noise level: 15
Intercept: -1.535
Coefficients:
 [[4.06304243]]
Explained variance score: 0.867
Mean squared error: 223.945
```

Figure 7-2 pictures how error accumulates; the line itself was calculated over the whole dataset, so it appears wrong on this subset. The MSE is simply an average of sum of squared errors (SSE). Each vertical dashed red line designates a single error term (difference between true y_i and its predicted value $\hat{y}_i$). The model's coefficients minimize this MSE (loss function), and the corresponding expression to calculate the vector β is an *unbiased minimal variance estimator*. It is independent of the inherent noise factor, and MSE is an estimator of this error term's spread ($\sigma \cong \sqrt{MSE}$). Estimates and values dependent upon them are denoted by a hat symbol.

269

Note   The intercept only makes sense when the matching predictor is of ratio level data type (the notion of absolute zero exists). Otherwise, it should be turned off (see the constructor for the LinearRegression class).

Hint   The print_parameters method is rudimentary. For a complete R-style summary of model parameters, you may want to utilize the statsmodels package (visit https://www. statsmodels.org).

Once a model is prepared (trained), it is ready for production usage to deal with unseen data. This is what I refer to as a *runtime model*. All learned (runtime) parameters are its integral part and are usually distributable over various communication channels. This property allows training to be done separately on powerful machines with lots of data. For example, all we need to know in production to handle real data is the corresponding vector $\hat{\beta}$. Calculating $\hat{y} = \hat{\beta}_0 + \hat{\beta}_{1:n}X$ can be performed on any constrained device.

While working with world1 (original model) we have used a trivial training process. All available data were reused both for training and testing, which is something you should avoid in practice. In this case, it didn't cause problems. Our model's complexity perfectly matched the truth, evidenced when higher noise steered a drop in explained variance as well as a rise of MSE. This was an indication that we haven't tried to capture nonessential properties of the real world. An overly complex model is capable of doing this, which leads to *overfitting*. Contrary to this is *underfitting*, when our model is too weak even for capturing fundamental characteristics. The next two sections demonstrate these aspects thoroughly.

$$SSE = \sum_{i=1}^{n}(y_i - \hat{y}_i)^2$$

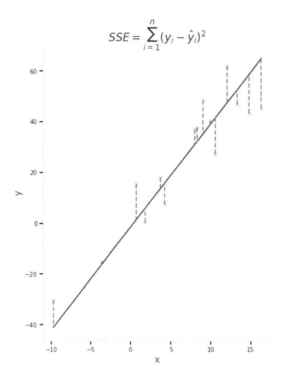

Figure 7-2. *Error terms* $e_i = y_i - \hat{y}_i$, *which are squared, summed up, and averaged to calculate MSE. The root MSE, ($\sqrt{MSE}$), is useful to restore the response's original unit.*

Overfitting

We will augment the linear model with polynomial features of various degrees; the model itself remains linear in terms of features. The aim is to demonstrate overfitting. Since this model is more powerful than the previously investigated model, it has enough capacity to encompass erroneous fluctuations in data; in a deterministic case (without error), you wouldn't notice any difference. Machine learning algorithms cannot decipher that irregularities in data aren't critical. There is a mechanism to detect overfitting by splitting historical data into training and testing sets.

Listing 7-4 shows the demo_overfitting function using polynomial features as well as the plot_mse function to plot MSE for both training and test sets of various sizes. During the run the data is split into training and test sets of varying sizes. This expansion of the training process introduces a new process parameter: data volume reserved for training purposes (the rest is kept for testing). Previously we just used whatever we had without any breakdown. The inner visualization function demonstrates what actually happens behind curtains.

Listing 7-4. Contains Functions to Exemplify Overfitting

```
def demo_overfitting():
    def visualize_overfitting():
        train_model(optimal_model, X, y)
        train_model(complex_model, X, y)

        _, ax = plt.subplots(figsize=(9, 7))
        ax.set_yticklabels([])
        ax.set_xticklabels([])
        ax.grid(False)

        X_test = np.linspace(0, 1.2, 100)
        plt.plot(X_test, np.sin(2 * np.pi * X_test),
        label='True function')
        plt.plot(
            X_test,
            optimal_model.predict(X_test[:, np.newaxis]),
            label='Optimal model',
            ls='-.')
        plt.plot(
            X_test,
            complex_model.predict(X_test[:, np.newaxis]),
            label='Complex model',
```

```
            ls='--',
            lw=2,
            color='red')
    plt.scatter(X, y, alpha=0.2, edgecolor='b', s=20,
    label='Training Samples')
    ax.fill_between(X_test, -2, 2, where=X_test > 1,
    hatch='/', alpha=0.05, color='black')
    plt.xlabel('x', fontsize=15)
    plt.ylabel('y', fontsize=15)
    plt.xlim((0, 1.2))
    plt.ylim((-2, 2))
    plt.legend(loc='upper left')
    plt.title('Visualization of How Overfitting Occurs',
    fontsize=19)
    plt.show()

set_session_seed(172)
X = generate_base_features(120)[['x_uniform']]
y = generate_response(X, 0.1, [0, 2 * np.pi], f=np.sin)

optimal_model = make_poly_pipeline(LinearRegression(), 5)
plot_mse(optimal_model, X, y, 'Optimal Model', 0.1)
complex_model = make_poly_pipeline(LinearRegression(), 35)
plot_mse(complex_model, X, y, 'Complex Model', 0.1)

visualize_overfitting()
```

When a model matches the truth, the MSEs induced by both the
training set and the test set gravitate around the achievable minimum MSE
(reflects inherent variance in data), as shown in Figure 7-3. Otherwise, the
test set shows a worse performance, as presented in Figure 7-4. This is a sign
that the model isn't generalizing properly and is picking up unimportant
details from a training set. As a model's power increases above the required
level, the cross-validation (CV) score considerably deteriorates, too.

The CV score is a very efficient way to evaluate your model's performance. It is an average of individual CV scores. The training data is randomly partitioned into K number of equal-sized complementary subsets (we use K=10; see the `collect_mse` inner function in Listing 7-2 as well as Exercise 7-2); this number K is another process-related parameter. In each round, one segment is used for testing (more precisely, for validation), while the rest is used as real training data (in the later section "Regularization," we will use *K-fold CV* to set a runtime parameter). Every data point from the initial dataset is used only once for testing. In this manner, the algorithm cycles through all partitions, so the overall mean score is indeed a reliable performance indicator. There are also extreme variants, where testing is only done with a single observation (singleton) per iteration (a.k.a. *leave-one-out CV*).

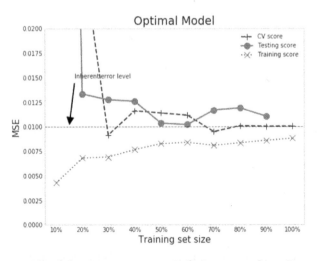

Figure 7-3. *MSE of the test set approaches natural variance when the model's complexity is just right*

Furthermore, the variance between CV, testing, and training scores lessens as we use more training data. There is no test score marker when we use all available data for training. The horizontal dashed line denotes the inherent error in data (it may stem from measurement errors). A proper model wouldn't try to embody this segment. In machine learning we cannot simply instruct the computer to forget about inherent error. The trick is to set your model's complexity just right, so that it has no capacity to memorize unwanted details. Apparently, the machine first tried to suck in everything, when the training set was small enough, but later gave up and only followed the main trend. Consequently, you must possess enough training data for a given model's complexity. Monstrous models (especially deep neural networks) must devour a huge amount of training data before being ready for production.

Overfitting and underfitting (demonstrated in the next section) are tightly associated with the central issue of supervised learning known as *bias-variance trade-off*. A highly biased model may miss important variances in the training data, while a high-variance model may try to capture nonessential properties. Balancing these two forces is one of the most difficult aspects of training machine learning algorithms. For example, extreme care must be given to decision trees, which may easily cover all edge cases in training data.

Figure 7-5 shows why an overly complex model doesn't generalize well; that is, it has low performance on a test set. Now, this set isn't exclusively about some totally uncharted area from a domain but is simply a collection of data points excluded from the training set. Reasoning about unknown space entails a different learning approach (see also reference [1]), which will be the topic of the next case study.

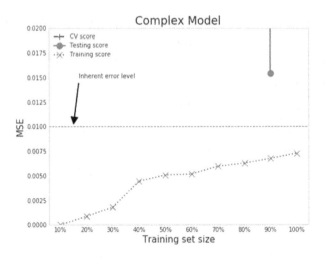

Figure 7-4. *MSE of the test set rises above the inherent error level when the model's complexity is too high*

The cross-validation score is even out of range here. Observe that the model completely memorized a small training sample at the beginning and always managed to follow unintended fluctuations. One way to combat this situation is to saturate the system with more training data. In our case, it has started to reasonably perform on a test set only from 60% of the training set size.

Underfitting and Feature Interaction

This section introduces another typical situation in which features interact. The goal is to showcase that it isn't enough to solely identify pertinent features. You must understand how they interrelate in the real world. We assume two models: one that treats individual features as independent, and another that includes their interaction.

Listing 7-5 shows the demo_underfitting function that demonstrates underfitting. When a model matches the truth, the MSEs induced by both the training set and the test set gravitate around the achievable minimum

MSE (similarly as shown in Figure 7-3). By contrast, Figure 7-6 illustrates what happens with a weak model; obviously, adding more data to a weak model doesn't help. Underfitting in practice is less common then overfitting, particularly with powerful deep neural networks.

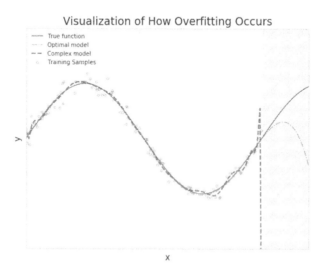

Figure 7-5. *The complex model's prediction line is jagged due to an attempt to encompass inherent error*

On the right side of Figure 7-5, you can see what happens in an unexplored territory. Both fitted lines depart from the true function.

Listing 7-5. Function to Exemplify Underfitting via Feature Interaction

```
def demo_underfitting():
    set_session_seed(15)
    X = generate_base_features(200)
    X_interacting = X[['x_interacting']]
    y = generate_response(X_interacting, 2, [1.7, -4.3])
```

```
plot_mse(LinearRegression(), X_interacting, y, 'Optimal
Model', 2)
X_weak = X[['x_normal', 'x_uniform']]
plot_mse(LinearRegression(), X_weak, y, 'Weak Model', 2)
```

The weak model properly enlists both participating features, but taken separately, they cannot provide value.

Figure 7-6. *MSEs of all sets are far away from inherent error with a weak model*

Collinearity

If you scrutinize the generate_base_features function, you will notice interrelationships between features x_normal, x_combined, and x_collinear. Our new simulated world uses only x_normal and x_combined. As before, we will presume two variants of this world: one using the same features, and the other one incorporating x_collinear, too. Listing 7-6 shows the code for demonstrating collinearity. In machine learning, this phenomenon has negative impact on performance and stability (it is hard to assess the impact of features on the outcome).

Listing 7-6. Extension of `session.py` with the `demo_collinearity`
Function

```
def demo_collinearity():
    set_session_seed(10)
    X = generate_base_features(1000)
    X_world = X[['x_normal', 'x_combined']]
    y = generate_response(X_world, 2, [1.1, -2.3, 3.1])

    model = LinearRegression()
    # Showcase the first assumed model.
    train_model(model, X_world, y)
    metrics = evaluate_model(model, X_world, y)
    print('\nDumping stats for model 1')
    print_parameters(model, metrics)

    # Showcase the second assumed model.
    X_extended_world = X[['x_normal', 'x_combined',
    'x_collinear']]
    train_model(model, X_extended_world, y)
    metrics = evaluate_model(model, X_extended_world, y)
    print('\nDumping stats for model 2')
    print_parameters(model, metrics)

    # Produce a scatter matrix plot.
    df = X
    df.columns = ['x' + str(i + 1) for i in range(len(df.
    columns))]
    df['y'] = y
    pd.plotting.scatter_matrix(df, alpha=0.2, figsize=(10, 10),
    diagonal='kde')
```

Here is the output of executing demo_collinearity():

```
Dumping stats for model 1
Intercept: 1.021
Coefficients:
 [[-2.1290165   3.05443636]]
Explained variance score: 0.999
Mean squared error: 4.274

Dumping stats for model 2
Intercept: 1.021
Coefficients:
 [[-2.1290165   0.09438926  0.52857984]]
Explained variance score: 0.999
Mean squared error: 4.274
```

The only difference between these runs is the significance of x_combined. This is a typical sign of collinearity. The system is confused whether to "work" with x_combined or x_collinear, since one is a direct linear combination of another. There is also a strong relation between x_normal and x_combined. A practical way to check for such interdependence is to generate a *scatter matrix plot*, as shown in Figure 7-7; this is the final result of running demo_collinearity (the columns are renamed to better fit on the diagram). It is possible to produce a correlation matrix, but it will only reveal linear relationships. A diagram may show you non-linear associations, too.

Independence assumption is common in machine learning algorithms. Naive Bayes method fully relies upon this characteristic. Feature dependence has negative consequences on convergence in logistic regression. Moreover, highly related features are redundant and just complicate a model.

Residuals Plot

As a data scientist, you must constantly seek to inspect problems from multiple angles. Likewise, you should know about complementary plotting techniques, since they could illuminate otherwise invisible aspects of a problem. You have just seen the utility of a scatter matrix plot. In this section we generate two worlds, one linear and one quadratic. Both have tiny coefficients, inherent randomness, and are approximated by truly linear models. Listing 7-7 contains the demo_residuals function to highlight the importance of residuals plots.

Figure 7-8 shows two scatter plots together with regression lines for fitting a linear world by linear model (case 1) and fitting a quadratic world by linear model (case 2). The MSE is same in both cases, while the explained variance score is better for case 2. Which case would you choose as more agreeable for a linear model? My guess is that you would pick case 2. Well, Figure 7-9 tells a different story.

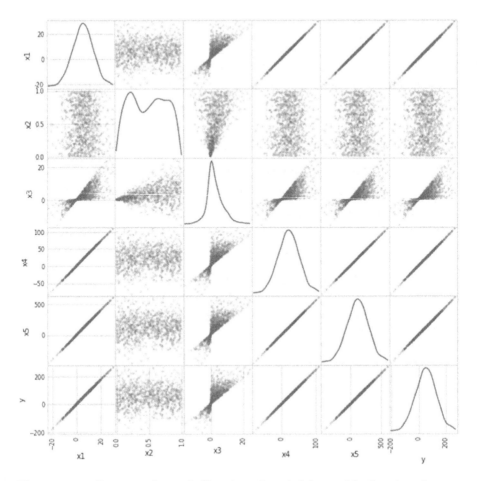

Figure 7-7. *Scatter plots of all pairs of variables, with density plots on diagonal*

Notice the heavy connection between x1, x4, x5, and y. Furthermore, observe x3's distinctive shape in relation to the other features.

Listing 7-7. Function That Creates Two Deceptive Offers, Where the Obvious One Isn't Obviously Wrong

```
def demo_residuals():
    def plot_regression_line(x, y, case_num):
        _, ax = plt.subplots(figsize=(9, 9))
```

```
    ax.set_title('Regression Plot - Case ' + str(case_num),
    fontsize=19)
    ax.set_xlabel('x', fontsize=15)
    ax.set_ylabel('y', fontsize=15)
    sns.regplot(x.squeeze(), y.squeeze(),
                ci=None,
                ax=ax,
                scatter_kws={'alpha': 0.3},
                line_kws={'color': 'green', 'lw': 3})

set_session_seed(100)
X = generate_base_features(1000)
X1 = X[['x_normal']]
y1 = generate_response(X1, 0.04, [1.2, 0.00003])
X2 = X1**2
y2 = generate_response(X2, 0.04, [1.2, 0.00003])

model = LinearRegression()
# Showcase the first world with a linearly assumed model.
plot_regression_line(X1, y1, 1)
train_model(model, X1, y1)
metrics = evaluate_model(model, X1, y1, True, 'Case 1')
print('\nDumping stats for case 1')
print_parameters(model, metrics)

# Showcase the second world with a linearly assumed model.
plot_regression_line(X1, y2, 2)
train_model(model, X1, y2)
metrics = evaluate_model(model, X1, y2, True, 'Case 2')
print('\nDumping stats for case 2')
print_parameters(model, metrics)
```

This code reuses the model instance from case 1 in case 2. You must always read carefully the documentation about what is happening when you call the fit method repeatedly. Here is the citation from scikit-learn's tutorial (see https://scikit-learn.org/stable/tutorial/basic/tutorial.html): "Hyper-parameters of an estimator can be updated after it has been constructed via the set_params() method. Calling fit() more than once will overwrite what was learned by any previous fit()." This behavior is exactly what we need.

You also must take care to reuse the same scaler instance that was used for training the model. The validation and test sets must be scaled in the same manner as the training dataset. Forgetting this detail may cause subtle bugs.

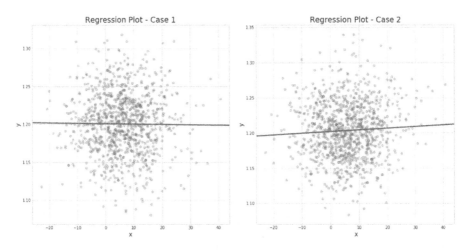

Figure 7-8. *The direction of the slope is wrong for case 1, since it should be positive (see Listing 7-7)*

For case 2 the slope is positive, as it should be. All signs suggest that this is a better match.

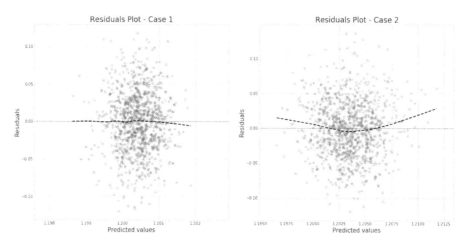

Figure 7-9. *The residuals plot clearly favors case 1 over case 2 with evident visual explanation; the curvature nicely reveals a "quadratic" pattern in residuals*

There is no residual pattern for case 1. See Listing 7-2 and the `evaluate_model` function (this creates a residuals plot with the lowest line for depicting trend).

Regularization

We don't know in advance which model is optimal. We can either start with a weak model and add more features or start with an overly complex model and try to tone it down. Neither of these approaches is scalable with manual work. The idea is to just err on the side of complexity and utilize some automation to reduce the model's complexity toward an optimal level. This is all about *regularization,* an automatic mechanism to eschew overfitting. There are many types of regularization. We will use the *Ridge regression* with built-in cross-validation.

Regularization encodes some constraint over coefficients using the language of mathematical optimization; this is expressed in the form of a *penalty function.* The Ridge regression (a.k.a. Tikhonov regularization) aims to keep coefficients as small as possible, since this is equivalent to attaining a least-complex model. Consequently, Ridge regression defines an L2 regularization term (l2-norm) $\alpha\|w\|_2^2$ that is added to the basic loss function (in our case MSE). The vector w contains the model's coefficients (weights). The parameter α balances minimization attempts between MSE and penalty term. Higher values tend to reward lesser coefficients, and vice versa. Alpha is calculated by some trial-and-error method (such as cross-validation over a list of candidates). Listing 7-8 shows the function to illuminate regularization, while the aftermath is shown in Figure 7-10.

Listing 7-8. Implementation of demo_regularization Function to Showcase Ridge Regression

```
def demo_regularization():
    from sklearn.linear_model import RidgeCV

    set_session_seed(172)
    x = generate_base_features(120)[['x_uniform']]
    y = generate_response(X, 0.1, [0, 2 * np.pi], f=np.sin)

    regularized_model = make_poly_pipeline(
            RidgeCV(alphas=[1e-3, 1e-2, 1e-1, 1, 5, 10, 20],
            gcv_mode='auto'),
            35)
    plot_mse(regularized_model, X, y, 'Regularized Model', 0.1)
```

All the magic happens inside the scikit-learn framework's RidgeCV class.

Predicting Financial Movements Case Study

The realm of financial modeling will shed some light on *time series analysis*, where we want to react to events in near real time (for example, predicting stock prices in markets). This is totally different than what we have done thus far using *batch processing.* My aim here isn't to develop a new breed of stock market application, but to drive your attention to novel problems and potential solutions with streaming data. You cannot access such data all at once, so *stream processing* techniques are intrinsically incremental and casual (they act on current knowledge to predict future outcomes). This entails that model parameters evolve over time instead of being fixed at the end of the training stage. Monitoring and regularizing this evolution are also some new tasks compared to classical approaches (see references [1-2]).

Another crucial difference is about timestamping of observations. In the Chapter 2 case study of e-commerce customer segmentation, the data was coarsely timestamped (data files were segregated by days). Here, each record will have its own absolute timestamp, so that we can monitor trend, seasonality, and other time-based patterns; relative timing isn't enough for this purpose. To avoid time zone–related difficulties, it is beneficial to register timestamps as seconds since the beginning of epoch time (or similar higher granularity scheme).

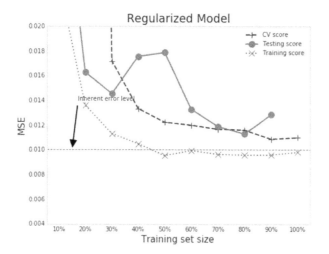

Figure 7-10. *Although the model's complexity, in term of features, is the same as in our example for overfitting, due to regularization it doesn't overfit as earlier*

Data Retrieval

The accompanying source code already contains the CSV file with daily time series stock data for Apple (its stock symbol is AAPL). Its located inside the stock_market subfolder (with other artifacts). You can get a fresh copy, or work with another company's equity (change the symbol below and rename the target file accordingly), by issuing the following command from Spyder's IPython console:

```
!curl -o daily_AAPL.csv ↪
"https://www.alphavantage.co/query?function=TIME_SERIES_DAILY&
symbol=AAPL&outputsize=full↪
&apikey=<YOUR API KEY>&datatype=csv"
```

We are relying on Alpha Vantage's API (see https://www.alphavantage.co/documentation/#daily) to retrieve daily time series data. You can get a free API key from Alpha Vantage and insert it into the URL in the preceding code where indicated. There are limits on the number of requests that are well documented on the site. The outputsize parameter is set to full, which pulls up to 20 years' worth of historical data. The output format (datatype parameter) is set to csv.

Data Preprocessing

In the spirit of the data science process, we will first do some preprocessing and exploratory data analysis. All subsequent steps should be carried out from Spyder's IPython console (ensure that you are in the stock_market folder). The next command shows the first five lines of the downloaded file:

```
>> !head -n 5 daily_AAPL.csv
timestamp,open,high,low,close,volume
2018-11-07,205.9700,210.0600,204.1300,209.9500,33106489
2018-11-06,201.9200,204.7200,201.6900,203.7700,31882881
```

```
2018-11-05,204.3000,204.3900,198.1700,201.5900,66163669
2018-11-02,209.5500,213.6500,205.4300,207.4800,91328654
```

According to the API documentation, "The most recent data point is the prices and volume information of the current trading day, updated realtime." We will omit this and work only with stable data points. The next lines read the stock data into a Pandas data frame and show the first couple of records:

```
>> import pandas as pd
>> stock_data = pd.read_csv('daily_AAPL.csv', usecols=[0, 4, 5],
   skiprows=[1])
>> stock_data.head()
    timestamp    close     volume
0   2018-11-06   203.77   31882881
1   2018-11-05   201.59   66163669
2   2018-11-02   207.48   91328654
3   2018-11-01   222.22   58323180
4   2018-10-31   218.86   38358933
```

We only need the timestamp, close price, and volume fields without the most recent data point. At this moment it is convenient to see the overall information about data types, number of rows, etc. The next command shows this information:

```
>> stock_data.info()
<class 'pandas.core.frame.DataFrame'>
RangeIndex: 5247 entries, 0 to 5246
Data columns (total 3 columns):
timestamp    5247 non-null object
close        5247 non-null float64
volume       5247 non-null int64
dtypes: float64(1), int64(1), object(1)
memory usage: 123.1+ KB
```

Apparently, the timestamp has an inconvenient object type, which is too generic to be useful. The next lines convert this column into a DateTime index (this allows us to treat data chronologically):

```
>> stock_data['timestamp'] = pd.to_datetime(stock_
   data['timestamp'])
>> stock_data.set_index('timestamp', inplace=True, verify_
   integrity=True)
>> stock_data.head()
            close     volume
timestamp
2018-11-06   203.77   31882881
2018-11-05   201.59   66163669
2018-11-02   207.48   91328654
2018-11-01   222.22   58323180
2018-10-31   218.86   38358933
```

Figure 7-11 shows AAPL closing levels over time, which is produced with the following snippet. The style parameter controls the appearance of lines in a plot. To differentiate lines in a grayscale image, you cannot solely use colors. We see a huge price drop around 2015 in Figure 7-11. According to one report (read at http://time.com/money/3991712/apple-stock-price-drop), the reason was a missed business expectation around iPhone sales.

```
>> import matplotlib.pyplot as plt

>> def plot_time_series(ts, title_prefix, style='b-'):
       ax = ts.plot(figsize=(9, 8), lw=2, fontsize=12,
       style=style)
       ax.set_title('%s Over Time' % title_prefix, fontsize=19)
       ax.set_xlabel('Year', fontsize=15)

>> plot_time_series(stock_data['close'], 'AAPL Closing Levels')
```

Discovering Trends in Time Series

The chart in Figure 7-11 is very ragged, due to noise and seasonality. A popular way to identify trends in time series is to take a *moving average*. The following command plots the trend in closing levels by calculating the *simple moving average (SMA)*, as shown in Figure 7-12:

```
>> stock_data.sort_index(inplace=True)
>> plot_time_series(stock_data['close'].rolling('365D').mean(),
   'AAPL Closing Trend')
```

If we don't sort the index, then we will receive an error, ValueError: index must be monotonic. The rolling window is defined to be 365 days. Figure 7-13 combines closing levels and volume trends inside a single diagram. The volume doesn't alter much over time, although when the price had started to drop then the volume had increased. Maybe there was an urge to sell stocks while the price was still good enough. This kind of comparison is useful for feature engineering and to get more insight into behavior. The next lines produce such a composite plot:

```
>> def compose_trends(ts):
       from sklearn.preprocessing import MinMaxScaler

       scaler = MinMaxScaler()
       scaled_ts = pd.DataFrame(scaler.fit_transform(ts),
       columns=ts.columns, index=ts.index)
       return pd.concat([scaled_ts['close'].rolling('365D').
       mean(),
                         scaled_ts['volume'].rolling('365D').
                         mean()], axis=1)

>> plot_time_series(compose_trends(stock_data), 'AAPL Closing &
Volume Trends', ['b-', 'g--'])
```

Taking a moving average eliminates small nuances in data. Also observe in Figure 7-13 that the slope of the massive price drop is less than in Figure 7-11. It is imperative to scale the features before combining them in the same diagram. Scaling is also mandatory in various machine learning situations, when movement in one direction may completely shadow movements in other directions. This happens when features are on totally different scales.

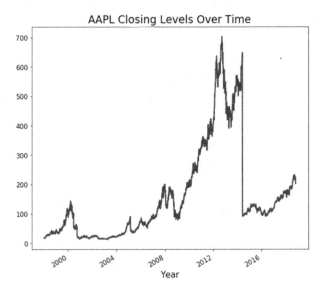

Figure 7-11. *Variation of AAPL closing levels over time*

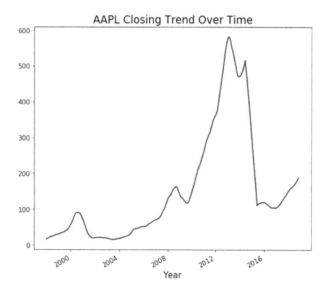

Figure 7-12. *Much smoother diagram compared to Figure 7-11*

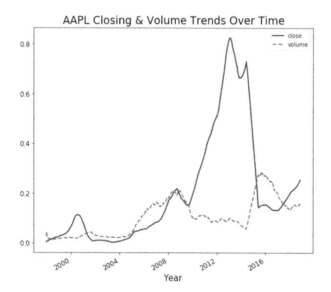

Figure 7-13. *The y axis is now showing scaled values*

Transforming Features

In finance, looking at returns on a daily basis is more useful than using absolute quantities. Returns indicate how much an asset's value fluctuates over time. There are two main ways to formulate returns:

- **Log returns:** $r_t = log\left(\dfrac{v_t}{v_{t-1}}\right)$, where v denotes the asset's value (such as closing price, adjusted closing price, volume, adjusted volume, etc.).

- **Scaled percent returns:** $r_t = \left(\dfrac{v_t}{v_{t-1}} - 1\right)$. There is a method pct_change in Pandas to calculate this quantity. For daily returns, scaled percent returns are near to log returns. You can see this from Taylor expansion of the log function, when $x = \dfrac{V_t}{V_{t-1}}$ is very small, so only the first term $(x - 1)$ matters.

Log returns has nice mathematical properties (such as additivity, symmetry pertaining to gains and losses, etc.). The following snippet attaches two new features to our data frame: the daily log returns of the stock price and the daily log changes in volume. Figure 7-14 shows AAPL price log returns over time, and Figure 7-15 presents the same for the volume.

```
>> import numpy as np

>> stock_data['close_ret'] = np.log(stock_data['close']).diff()
>> stock_data['volume_ret'] = np.log(stock_data['volume']).diff()
>> stock_data.head()
            close    volume   close_ret   volume_ret
timestamp
```

1998-01-02	16.25	6411700	**NaN**	**NaN**
1998-01-05	15.88	5820300	-0.023032	-0.096773
1998-01-06	18.94	16182800	0.176216	1.022597
1998-01-07	17.50	9300200	-0.079075	-0.553913
1998-01-08	18.19	6910900	0.038671	-0.296936

```
>> stock_data.dropna();
>> plot_time_series(stock_data['close_ret'], 'AAPL Price Log
   Returns')
>> plot_time_series(stock_data['volume_ret'], 'AAPL Volume Log
   Returns')
```

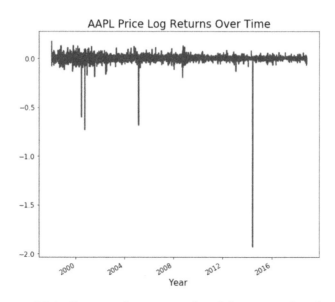

Figure 7-14. *This diagram has a couple of downward spikes that are far away from central values*

We could treat spikes as outliers, which are hard to model and account for (unless you are a finance guru). One way to circumvent this problem is to use volatility-normalized log returns.

The next code section implements the logic to produce volatility-normalized log returns of the stock price, as shown in Figure 7-16. There is no need to perform the same for the volume returns, as we will soon see (its distribution is nearly normal).

```
>> ts['close_ret'] /= ts['close_ret'].ewm(halflife=halflife).std()
>> plot_time_series(stock_data['close_ret'], 'AAPL Volatility-
   Norm. Price Log Returns')
```

The half-life of 23 days amounts to a decay factor (weight) of 0.97 (you need to solve for w the equation $w^{23} = \frac{1}{2}$), which controls how long the stock market remembers (or how fast it forgets) old events. The i-th data point has a decay factor of w^i. Volatility is calculated as a rolling standard deviation across data points taking into account the exponential decay for old data. Higher decay damps volatility. You should experiment with this factor to match the desired risk level.

Streaming Amounts

For simplicity reasons, we have thus far managed the data in batch mode. Nonetheless, log returns, moving average, and volatility are all amounts that may be calculated in real time. For log returns, you just need to cache the last value to make an update. The same is true for volatility with exponential downweighting, although it is not as obvious as with log returns. To track a moving average, you will need to store and update the last numerator and denumerator.

Suppose that you know the current weighted variance $V_{current} = \frac{\sum_i w^i r_i^2}{\sum_i w^i}$, where w is the decay factor; for daily returns, we may assume the mean to be zero. The denominator is the sum of a geometric series whose limit is $\dfrac{1}{1-w}$. When a new return r_0 arrives, then we have

$$V_{new} = (1-w)\left(w\sum_i w^i r_i^2 + r_0\right) = wV_{current} + (1-w)r_0.$$ You can compute the

current volatility as $\sqrt{V_{current}}$.

To perform volatility normalization in streaming mode, you need

to divide the new return with the current volatility: $r_0 \leftarrow \dfrac{r_0}{\sqrt{V_{current}}}$. Of

course, you would do this before updating the current volatility; that is,
before executing $V_{current} \leftarrow V_{new}$.

Besides computing various running totals, there is a whole gamut of
incremental algorithms that may run in streaming fashion. For example,
gradient descent is an iterative optimization method that handles all data in
one sweep. *Stochastic gradient descent* is an incremental and iterative method
that runs in online mode and updates parameters on-the-fly. Streaming
linear regression uses this approach (as described later in this chapter).

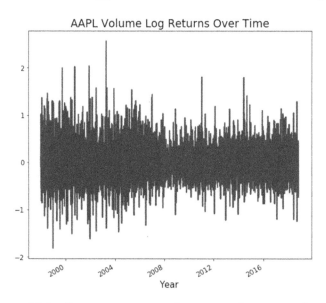

Figure 7-15. *This diagram seems to be properly centered around
zero, but we should still eyeball its distribution*

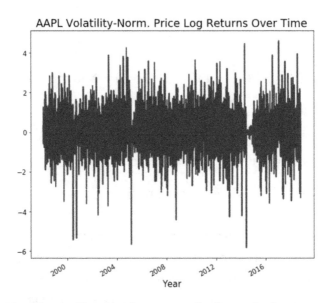

Figure 7-16. *Normalization has smoothed out the log returns and made the time series better behaved*

To get a better feeling of what normalization did to price log returns, Figures 7-17 and 7-18 show the histograms of non-normalized and normalized variants, respectively. Later in the section you will see the complete code.

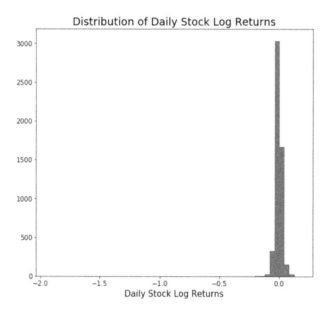

Figure 7-17. *Histogram of the non-normalized price log returns, which is heavily left-tailed*

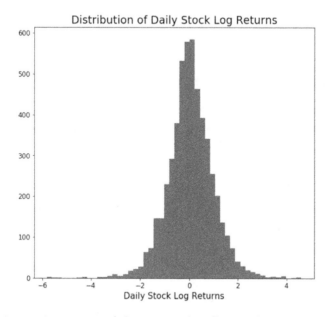

Figure 7-18. *Histogram of the normalized price log returns, which is bell shaped*

For the sake of completeness, Figure 7-19 shows the histogram of the volume log returns.

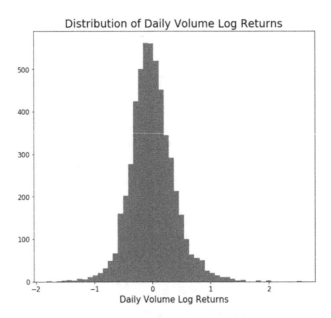

Figure 7-19. *This diagram is normally distributed with a slight right tail*

Listings 7-9, 7-10, and 7-11 show separate modules that bundle all pertinent steps into coherent units. This concludes the preprocessing stage. The driver.py module calls these functions to implement the whole pipeline.

Listing 7-9. data_preprocessing.py Module, Which Contains Functions to Prepare the Data Frame for the Analysis Phase

```
import numpy as np
import pandas as pd

def read_daily_equity_data(file):
    stock_data = pd.read_csv(file, usecols=[0, 4, 5], skiprows=[1])
```

```python
    stock_data['timestamp'] = pd.to_datetime(stock_
    data['timestamp'])
    stock_data.set_index('timestamp', inplace=True, verify_
    integrity=True)
    stock_data.sort_index(inplace=True)
    return stock_data

def compose_trends(ts):
    from sklearn.preprocessing import MinMaxScaler

    scaler = MinMaxScaler()
    scaled_ts = pd.DataFrame(scaler.fit_transform(ts),
    columns=ts.columns, index=ts.index)
    return pd.concat([scaled_ts['close'].rolling('365D').mean(),
                      scaled_ts['volume'].rolling('365D').
                      mean()], axis=1)

def create_log_returns(ts, halflife, normalize_close=True):
    ts['close_ret'] = np.log(ts['close']).diff()
    if normalize_close:
        ts['close_ret'] /= ts['close_ret'].
        ewm(halflife=halflife).std()
    ts['volume_ret'] = np.log(ts['volume']).diff()
    return ts.dropna()
```

Listing 7-10. data_visualization.py Module, Which Contains
Auxiliary Visualizations of Time Series

```python
import matplotlib.pyplot as plt

def plot_time_series(ts, title_prefix, style='b-'):
    ax = ts.plot(figsize=(9, 8), lw=2, fontsize=12,
    style=style)
    ax.set_title('%s Over Time' % title_prefix, fontsize=19)
```

```
    ax.set_xlabel('Year', fontsize=15)
    plt.show()

def hist_time_series(ts, xlabel, bins):
    ax = ts.hist(figsize=(9, 8), xlabelsize=12, ylabelsize=12,
    bins=bins, grid=False)
    ax.set_title('Distribution of %s' % xlabel, fontsize=19)
    ax.set_xlabel(xlabel, fontsize=15)
    plt.show()
```

Listing 7-11. `driver.py` Module, Which Connects All the Pieces Together (First Part of File Shown)

```
from data_preprocessing import *
from data_visualization import *

stock_data = read_daily_equity_data('daily_AAPL.csv')
stock_data = create_log_returns(stock_data, 23)

plot_time_series(stock_data['close'], 'AAPL Closing Levels')
plot_time_series(stock_data['close'].rolling('365D').mean(),
'AAPL Closing Trend')
plot_time_series(compose_trends(stock_data), 'AAPL Closing &
Volume Trends', ['b-', 'g--'])

# To produce the non-normalized price log returns plot you must
call
# the create_log_returns function with normalize_close=False.
Try this as an
# additional exercise.
plot_time_series(stock_data['close_ret'], 'AAPL Volatility-
Norm. Price Log Returns')
plot_time_series(stock_data['volume_ret'], 'AAPL Volume Log
Returns')
```

```
hist_time_series(stock_data['close_ret'], 'Daily Stock Log
Returns', 50)
hist_time_series(stock_data['volume_ret'], 'Daily Volume Log
Returns', 50)
```

Feature Engineering

Currently, we have raw close levels, raw volume levels, volatility-normalized closing log returns, and volume log returns as our features (we will create more). To see how these impacts a potential target (like, predicted stock price change) it is useful to consult Pearson's correlation coefficient r. It is an indicator for a linear relationship between two features, whose range is [-1, 1]. A positive correlation means as one value increases/decreases the other does the same. A negative correlation denotes the opposite behavior. A value of zero represents the absence of a linear relationship, although the quantities may be interrelated in non-linear ways. Usually, if $|r| > 0.3$, then we can consider the correlation to be noticeable (this a judgement call, so take this heuristics with a decent pinch of salt).

Log returns denote fluctuations and it is illuminating to find out whether these are *mean-reverting* or *trend-following* in some time period (for example, N days). Mean reversion suggests that returns oscillate around a mean, while trend-following says that they mimic the recent period. We can discover this by fixing N and calculating the coefficient r between returns from past and future. If the correlation is low, then we have mean reversion, otherwise trend-following behavior. Listing 7-12 shows the function that reports correlation coefficients and creates scatter plots between past and future price as well as volume log returns (see also Exercise 7-3). It calls the `scatter_time_series` function to make a scatter plot (see Listing 7-13).

Listing 7-12. `feature_engineering.py` Module, with Function to Investigate Auto-correlation

```
from data_visualization import *

def report_auto_correlation(ts, periods=5):
    for column in filter(lambda str: str.endswith('_ret'),
    ts.columns):
        future_column = 'future_' + column
        ts[future_column] = ts[column].shift(-periods).
        rolling(periods).sum()
        current_column = 'current_' + column
        ts[current_column] = ts[column].rolling(periods).sum()

        print(ts[[current_column, future_column]].corr())
        scatter_time_series(ts, current_column, future_column)
```

Listing 7-13. Function to Create a Scatter Plot to Trace Auto-correlation (`data_visualization.py` Module)

```
def scatter_time_series(ts, x, y):
    ax = ts.plot(x=x, y=y, figsize=(9, 8), kind='scatter',
    fontsize=12)
    ax.set_title('Auto-correlation Graph', fontsize=19)
    ax.set_xlabel(x, fontsize=15)
    ax.set_ylabel(y, fontsize=15)
    plt.show()
```

The following are the correlations for both quantities, and Figure 7-20 shows the scatter plot for volume log returns (you should run the accompanying source code to see the graph for price returns):

	current_close_ret	future_close_ret
current_close_ret	1.000000	**0.013661**
future_close_ret	**0.013661**	1.000000

	current_volume_ret	future_volume_ret
current_volume_ret	1.000000	**-0.419489**
future_volume_ret	**-0.419489**	1.000000

We may conclude that price log returns revert to the mean, while volume log returns follow the opposite trend.

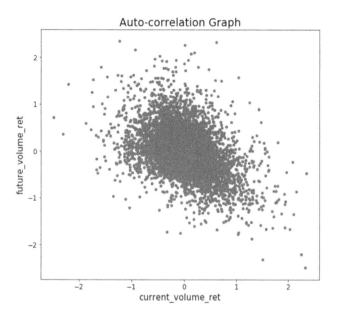

Figure 7-20. *As current volume log returns increase, future volume log returns decrease, and vice versa*

By varying the period's length, it is possible to find the highest level of auto-correlation, which may serve as a basis for creating features. It seems that 5 days is a good choice (you may experiment with different periods using the accompanying code base). Therefore, out target feature will be the 5-days future price change, while our initial basic features will be the current 5-days price and volume changes.

Every domain has its own set of prepared favorite features that have been proven to be valuable in predicting targets. There is a powerful technical analysis library with 200 financial indicators called TA-Lib (see https://mrjbq7.github.io/ta-lib). We will use three of them: normalized SMA, relative strength index (RSI), and on-balance volume (OBV). You have already seen SMA in the "Discovering Trends in Time Series" section. RSI is defined as $100 - \dfrac{100}{1 + RS}$, where $RS = \dfrac{mean\ gain\ over\ a\ period}{mean\ loss\ over\ a\ period}$. OBV is connecting volume flow with price changes. Listing 7-14 contains the code for creating candidate features.

Listing 7-14. Code to Create Features Described in Text (Part of feature_engineering.py)

```
def create_features(ts):
    from talib import SMA, RSI, OBV

    target = 'future_close_ret'
    features = ['current_close_ret', 'current_volume_ret']

    for n in [14, 25, 50, 100]:
        ts['sma_' + str(n)] = SMA(ts['close'].values,
        timeperiod=n) / ts['close']
        ts['rsi_' + str(n)] = RSI(ts['close'].values,
        timeperiod=n)
    ts['obv'] = OBV(ts['close'].values, ts['volume'].
    values.astype('float64'))

    ts.drop(['close', 'volume', 'close_ret', 'volume_ret',
    'future_volume_ret'],
            axis='columns',
            inplace=True)
    ts.dropna(inplace=True)
    return ts.corr()
```

The function returns the correlation matrix, which may be conveniently visualized using a heat map. It is also possible to return an exponentially weighted correlation matrix and to use a weighted moving average (see the Pandas EWM object's corr and mean methods, respectively). The heat map plotting is implemented in a new function, as shown in Listing 7-15 (inside our visualization module).

Listing 7-15. Code to Report Correlations Using Seaborn's heatmap Facility

```python
def heat_corr_plot(corr_matrix):
    import numpy as np
    import seaborn as sns

    mask = np.zeros_like(corr_matrix)
    mask[np.triu_indices_from(mask)] = True
    _, ax = plt.subplots(figsize=(9, 8))
    sns.heatmap(corr_matrix, annot=True, cmap='gist_gray',
    fmt=".2f", lw=.5, mask=mask, ax=ax)
    plt.tight_layout()
    plt.show()
```

Finally, the driver.py module must be extended with the following statements:

```python
from feature_engineering import *

report_auto_correlation(stock_data)
corr_matrix = create_features(stock_data)
heat_corr_plot(corr_matrix)
```

Figure 7-21 shows the correlation matrix, which can be used for filtering features. Reducing the number of features may improve both accuracy and performance.

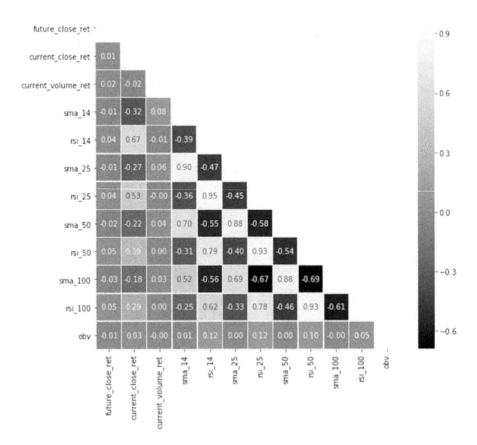

Figure 7-21. *This diagram shows only the relevant part of the correlation matrix*

All diagonal elements are 1, since features are perfectly aligned with themselves. The matrix is symmetric, so the upper-right triangle is redundant. None of the features are strongly coupled with the target, at least not in a linear way.

Implementing Streaming Linear Regression

Streaming linear regression continually learns and updates its parameters as new training data is added to the corpus. We rely on Apache Spark's MLlib framework that provides the `StreamingLinearRegressionWithSGD` class to perform streaming regression. The only task for us is to provide streaming sources for training and test data. One handy way is to deliver a list of *resilient distributed dataset (RDD)* instances over a queue. Converting a section of the Pandas DataFrame into an RDD is straightforward. Listing 7-16 shows the implementation to fit a linear model in online regime (see also Exercise 7-4), and Listing 7-17 shows the final expansion of the driver.py module. You will need to install pyspark (see https://spark.apache.org/docs/latest/api/python/index.html).

Listing 7-16. Content of the streaming_regression.py Module to Showcase Streaming Linear Regression in Online Mode

```python
def fit_and_predict(sparkSession, ts):
    import numpy as np
    from sklearn.model_selection import train_test_split
    from pyspark.streaming import StreamingContext
    from pyspark.mllib.regression import
    StreamingLinearRegressionWithSGD

    def to_scaled_rdd(pandasDataFrame):
        import pandas as pd
        from sklearn.preprocessing import RobustScaler
        from pyspark.mllib.regression import LabeledPoint

        regressors = pandasDataFrame.columns[1:]
        num_regressors = len(regressors)
        # FIX ME: As a bonus exercise, read the last paragraph
        from section about residual
        # plots and make the necessary bug fix! Compare the
        behavior of this version with the
        # fixed one and see whether you can decipher anything
        from the outputs.
        scaler = RobustScaler()
```

```
        scaled_regressors = scaler.fit_transform(pandasDataFram
        e[regressors])
        scaled_pandasDataFrame = pd.DataFrame(scaled_
        regressors, columns=regressors)
        scaled_pandasDataFrame['target'] =
        pandasDataFrame[pandasDataFrame.columns[0]].values

        sparkDataFrame = sparkSession.createDataFrame(scaled_
        pandasDataFrame)
        return sparkDataFrame.rdd.map(
                lambda row: LabeledPoint(row[num_regressors],
                row[:num_regressors]))
    def report_accuracy(result_rdd):
        from pyspark.mllib.evaluation import RegressionMetrics

        if not result_rdd.isEmpty():
            metrics = RegressionMetrics(
                    result_rdd.map(lambda t: (float(t[1]),
                    float(t[0]))))
            print("MSE = %s" % metrics.meanSquaredError)
            print("RMSE = %s" % metrics.rootMeanSquaredError)
            print("R-squared = %s" % metrics.r2)
            print("MAE = %s" % metrics.meanAbsoluteError)
            print("Explained variance = %s" %
            metrics.explainedVariance)

    df_train, df_test = train_test_split(ts, test_size=0.2,
    shuffle=False)
    train_rdd = to_scaled_rdd(df_train)
    test_rdd = to_scaled_rdd(df_test)

    streamContext = StreamingContext(sparkSession.sparkContext, 1)
    train_stream = streamContext.queueStream([train_rdd])
    test_stream = streamContext.queueStream([test_rdd])
```

```
numFeatures = len(ts.columns) - 1
model = StreamingLinearRegressionWithSGD(stepSize=0.05,
numIterations=300)
np.random.seed(0)
model.setInitialWeights(np.random.rand(numFeatures))

model.trainOn(train_stream)
result_stream = model.predictOnValues(test_stream.
map(lambda lp: (lp.label, lp.features)))
result_stream.cache()
result_stream.foreachRDD(report_accuracy)

streamContext.start()
streamContext.awaitTermination()
```

Listing 7-17. Last Piece of the Main driver.py Module

```
from pyspark.sql import SparkSession

from streaming_regression import *

sparkSession = SparkSession.builder \
                        .master("local[4]") \
                        .appName("Streaming Regression Case
                        Study")\
                        .getOrCreate()
fit_and_predict(sparkSession, stock_data)
```

The system will print the following report (after reading this, you should terminate the session):

```
MSE = 5.663621710391528
RMSE = 2.379836488162901
R-squared = -0.3217279555407153
MAE = 1.8382286514438086
Explained variance = 1.7169768418351132
```

At the time of this writing, the `StreamingLinearRegressionWithSGD` class is missing the `setIntercept` method. Consequently, we have very "strange" values for R-squared and explained variance. It is also crucial to remember the importance of scaling features before training the model. The `RobustScaler` scaler is convenient if you aren't sure about the distribution of your regressors.

EXERCISE 7-1. IMPROVE REUSABILITY

The `plot_mse` function presets font sizes for title and axes. This tactic is also repeated in other places to preserve consistency. Nonetheless, this approach is a real maintenance nightmare (notice that `explain_sse` repeats the same setup). Improve the code to centralize setting common parameters (hint: read about style sheets at `https://matplotlib.org/users/customizing.html`). What else can you devise to make the code base more flexible and reusable?

As you work on various machine learning problems, utility functions become very handy. There is no point in reinventing the wheel each time you need to plot MSEs. Try to make your code base reusable as much as you can.

EXERCISE 7-2. FIX A BUG

In Listing 7-2, inside the `plot_mse` function, you will find the following two lines of code:

```
# FIX ME: See Exercise 2!
ax.set_ylim(max(0, y_min), min(2 * error_variance, y_max))
```

Your task is to find out why this marked line is wrong and implement a fix. In reality, the situation is even worse, because nobody will point out to you a wrong section of code. Locating the exact place of an issue is a huge milestone toward correcting errors.

Hint: Figure 7-6 was produced with both lines commented out. Obviously, such resolution is a quick and dirty hack. Try running demo_underfitting with the original plot_mse function and see what happens.

EXERCISE 7-3. AVOID SIDE-EFFECTS

In Listing 7-12 you can see a function with side effects. It modifies the input data frame with extra columns. This modification is a prerequisite for calling the function presented in Listing 7-14. All the chaining is driven from the driver module.

Programming in terms of side effects is generally a bad practice. You may wonder why the model is altered in this fashion (see Listing 7-2) after calling its fit method. The crucial difference is that model encapsulates its state inside an object, where you have more control over changes to internal stuff. When ordinary functions are spread out with assumed side effects, then it is easy to lose control and create an unmaintainable mess.

Refactor the feature_engineering.py module to group feature-related manipulations inside a dedicated class. Compare how such object orientation helps to retain control over internals of the system.

EXERCISE 7-4. EXPERIENCE STREAMING BEHAVIOR

In Listing 7-16 the training and test streams were constructed in the following fashion:

```
train_stream = streamContext.queueStream([train_rdd])
test_stream = streamContext.queueStream([test_rdd])
```

This was OK to demonstrate the scaffolding of the solution but makes no sense in a real environment. The whole point of streaming is to allow data to continuously arrive in chunks. Apache Spark even allows streams to be

313

combined, so that you may have data coming over multiple channels (for example, you can also monitor a folder for new data files and parse them in real time).

Refactor the solution to have many parts in the training and test sets. You need to split the df_train and df_test referenced data frames into sections and convert each into an RDD. Finally, provide these parts in a list to the queueStream method. Observe what you get on the output (you should receive as many reports as there are pieces of test data).

Bonus task: The current accuracy report doesn't reveal too much about performance. You may want to create a scatter plot of predictions vs. actual values (observations from test data). Draw also the ideal line (this is basically the function y=x).

Summary

This chapter barely scratches the surface of the machine learning (ML) domain. Even dozens of books wouldn't be enough to fully cover (even at an introductory level) all available algorithms and technologies. The major aim of this chapter was to present common concepts that permeate the ML knowledge area. Without being aware of underfitting, overfitting, regularization, scaling, and similar topics, there is no way to be efficient with any ML approach. To be proficient in ML, you must also recall the golden tenet of data science and engineering: "Keep it simple, stupid!" (a.k.a. KISS principle). Typically, you don't even need ML to solve a problem, and rarely will you ever need to fire up complicated deep neural networks.

Another potent message of this chapter is the rather blurred borderline between science and art regarding parameter tuning. There are some rules of thumb (for a good overview, I suggest the document at http://martin.zinkevich.org/rules_of_ml/rules_of_ml.pdf), but you will need lots of experimentation and trial. Consequently, to make all this happen in reasonable amount of time, you will need powerful hardware (readily available in a cloud as infrastructure or platform as a service).

ML is all about mathematics. There is no way to escape this fact. Neural networks appear to give you a seemingly good escape route (we will cover them in detail in Chapter 12), but eventually you will need to dig under the hood to understand what is going on. This is tightly associated with interpretability of your model. With our ordinary least-square method of regression, coefficients can be easily explained. A particular coefficient represents how much the target changes for a unit change in the corresponding feature. With more complex models, the situation is different. At any rate, I will talk more about inner characteristics of models in Chapter 9.

ML is getting huge attention from both research institutions and industry. This isn't surprising, since as we delve more into the realm of Big Data, there is a greater need to handle such large amounts of data in an efficient manner. Only with the help of machines can we hope to seize control of massive data. One hot topic is *transfer learning*, which is an attempt to boost reuse of trained models. The idea is to leverage models optimized for one task to perform well on other tasks, too (maybe with minor extra tweaking and training). This will surely trigger a whole bunch of new algorithms and technologies.

One urgent need in the area of machine learning is an efficient and standardized way to exchange prediction models. One such standard is the Predictive Model Markup Language (PMML). There is a Python library for converting scikit-learn pipelines to PMML at `https://github.com/jpmml/sklearn2pmml`.

References

1. Christopher Bishop, *Pattern Recognition and Machine Learning*, Springer, 2006.

2. Nathan George, "Machine Learning for Finance in Python," DataCamp, `https://www.datacamp.com/courses/machine-learning-for-finance-in-python`.

3. Ian H. Witten, Eibe Frank, Mark A. Hall, and Christopher J. Pal, *Data Mining: Practical Machine Learning Tools and Techniques, Fourth Edition*, Morgan Kaufmann, 2016.

CHAPTER 8

Recommender Systems

When someone asks where machine learning is broadly applied in an industrial context, recommender systems is a typical answer. Indeed, these systems are ubiquitous, and we rely on them a lot. Amazon is maybe the best example of an e-commerce site that utilizes many types of recommender systems to enhance users' experience and help them quickly find what they are looking for. Spotify, whose domain is music, is another good example. Despite heavy usage of machine learning, recommender systems differ in two crucial ways from classically trained ML systems:

- Besides basic content, they try to combine many additional clues to offer more personalized choices.

- They tend to favor variability over predictability to boost excitement in users. Just imagine a recommender system whose proposals would never contain surprising items.

These traits entail a completely different set of metrics to evaluate recommender systems. If you were to try to optimize prediction accuracy solely by using the standard root mean squared error, then your system would probably not perform well in practice; it would be judged as boring. This chapter briefly introduces you to recommender systems and explains some core concepts and related techniques.

© Ervin Varga 2019
E. Varga, *Practical Data Science with Python 3*,
https://doi.org/10.1007/978-1-4842-4859-1_8

Introduction to Recommender Systems

To familiarize yourself with a recommender system, try out the freely available MovieLens project (`https://movielens.org`) without any commercial baggage.[1] MovieLens provides personalized movie recommendations. After creating an account, it immediately asks you to distribute three points among six possible genres of movies to produce an initial profile. Later, as you rate and tag movies, it learns more about you and offers better-suited movies. This is a typical case in recommender systems: more data allows the system to create a finer-grained profile about you that can be used to filter content more successfully. In this respect, high-quality input should result in high-quality output. Figure 8-1 shows part of the main user interface of MovieLens. Try to rate a couple of movies and watch how the system adapts to your updated profile.

As with MovieLens, most recommender systems collect two types of data:

- **Explicit data**: Voluntarily given by users. For example, each time you rate a movie, you intentionally provide feedback. Collecting lots of explicit data requires system designers to find ingenious ways to entice users to respond, such as by offering incentives. Of course, if you properly rate movies, then you should get better recommendations, which is the minimal incentive for anyone to bother rating products.

- **Implicit data**: Collected automatically by a system while monitoring activities of users (such as clicking, purchasing, etc.). Even timing events may be valuable; for example,

---

[1]MovieLens is part of the GroupLens research project (see `https://grouplens.org`), which allows you to download free datasets composed of anonymized movie ratings. We will use some of this data in subsequent examples (see reference [1]). Another associated project is LensKit (see `https://lenskit.org`) for building experimental recommender systems in Python (some earlier editions were Java based). You will see examples using this framework.

if a user has spent more time on a page showing the script of a particular movie than she has spent on the pages of other movies, that may indicate a higher interest in that movie. Another possibility for collecting implicit data is to trace tweets about movies to evaluate their general popularity. All in all, a multitude of sources could be combined into a single preference formula. Naturally, acquiring implicit data is easier than acquiring explicit data, as it doesn't require active participation by users.

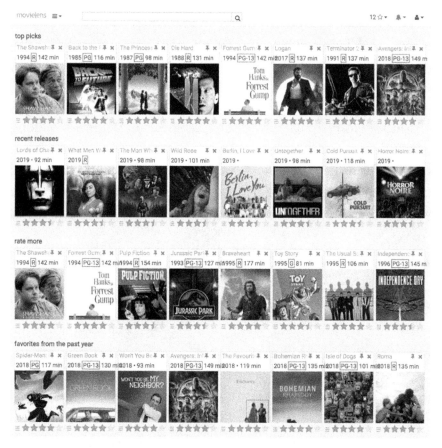

Figure 8-1. *MovieLens recommendations are displayed in rows that denote various categories*

To always give users a chance to discover new stuff, recommendations are a mixture of personalized content and nonpersonalized content (such as overall popular items). The latter may be customized based on general demographic information (for example, age group, gender, etc.). Another consideration is that sometimes you just want to explore things without "disturbing" your profile (remember that every action is tracked, including browsing and clicking, which might influence future offerings). In MovieLens, this is possible via the MovieExplorer feature (it may change over time, as it is purely tentative at the time of this writing). At any rate, consider in your design a similar possibility to offer an "untracked" option for your customers.

Figure 8-2 highlights another peculiarity of recommender systems compared to classical databases and other information retrieval systems. Recommender systems filter items based on learned preferences, while a database tries to answer ad hoc queries as fast as possible. The set of queries executed by MovieLens is pretty static (for example, list top movies for a user, list most popular movies, etc.), as shown in Figure 8-1, although you get different results. If you run the same query against a relational database system, then you will get back the same output (assuming that the basic content didn't change) irrespective of whether you like it or not.

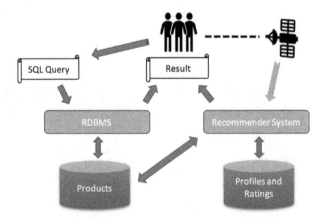

Figure 8-2. *Within an information retrieval system, a user must explicitly identify what she is looking for (for example, in a relational database, this is accomplished by issuing SQL queries)*

A recommender system learns what a user likes and dislikes and modifies the underlying queries accordingly. The output reflects the current profile (i.e., the outcome resonates with the user's taste, assuming a personalized solution). Typically, you don't often alter the output categories in a recommender system; they are quite stable over time. A content-based recommender system may use the products database to display details about items, so it can encode taste in terms of attributes of items.

There are three major interfaces pertaining to recommender systems:

- **Filtering facilities**: Try to remove uninteresting topics (according to the user profile) from a streaming data source, such as news, tweets, e-mails, etc. For example, a personalized news suggestion engine may pick only articles that could be significant to you.

- **Recommendation interfaces**: Present a selection of items that you may like, such as shown earlier in Figure 8-1. Usually, modern recommenders are hybrid systems that leverage different recommendation algorithms to produce a mix of content. For example, it can give you both specific items matching your taste as well as suggestions for items that are currently popular. Furthermore, the system may only use its own initial recommendation to trigger a dialog-based interaction, where additional feedback from a user may help narrow down the final selection.

- **Prediction interfaces**: Attempt to accurately foresee how much you would like some products (for example, the number of stars you would assign to a movie you haven't viewed). Of course, these values are estimates and may sometimes turn out to be inaccurate. In general, it doesn't matter that much whether you will give 5 stars or 4.5 stars to an item. It is more important to offer things that you will find valuable.

Nowadays, most recommender systems are based on collaborative filtering techniques. Collaboration means that data from other users is leveraged to craft offerings for you. A system may correlate your taste with other users, so that you get recommendations based on what those users liked and disliked. This is the *user-user* type of filtering. It is also possible to correlate items regarding their ratings, which constitutes the *item-item* type of filtering; this is the preferred scalable model in most systems today. Finally, it is also possible to recommend items using product association rules of the form "users who bought this also bought..."

Context-based systems also use contextual and environmental data to fine-tune the result (for example, depending on your mood, the system may offer different music on top of your general preference model). Such data may be collected from your smartphone; the system may use location information to reduce the list only to objects that are in your vicinity.

Simple Movie Recommender Case Study

We will build a very simple mashup to recommend movies similar to a movie entered by a user. The inspiration for this example comes from reference [3]. The program will use two public services and combine their result; this arrangement is known as a *mashup*.

TasteDive (`https://tastedive.com`, which is also its API's base URL) is a recommendation engine for suggesting similar music, movies, TV shows, books, authors, games, and podcasts based on your taste. It offers a Resource API (level 2 REST) to acquire similar artifacts in programmatic fashion (the API is documented at `/read/api`). If you just hit the `/api/similar` endpoint (without entering an access key or providing any other parameters), you will receive the following abbreviated JSON response (the content may change over time):

```
{
  "Similar": {
    "Info": [
      {
        "Name": "!!!",
        "Type": "music"
      }
    ],
    "Results": [
      {
        "Name": "Meeting Of Important People",
        "Type": "music"
      },
      {
        "Name": "The Vanity Project",
        "Type": "music"
      },
      ...
    ]
  }
}
```

You can issue a couple of HTTP requests for free. You will need an access key to put your future recommender service into production. Listing 8-1 shows the class to retrieve artifacts from TasteDive. This program is totally independent from our final service. It can be reused in many other contexts, too. As an additional exercise, change the requests module to requests_with_caching (look it up in Chapter 24 of reference [3]).

Listing 8-1. `tastedrive_service.py` Module in the `simple_` `recommender` Folder for Getting Similar Stuff from TasteDive

```python
import requests

class TasteDiveService:
    SUPPORTED_ARTIFACTS = ['music', 'movies', 'shows',
    'podcasts', 'books', 'authors', 'games']
    API_URL = 'https://tastedive.com/api/similar'

    def __init__(self, artifact_type = 'movies'):
        assert artifact_type in TasteDiveService.SUPPORTED_
        ARTIFACTS, 'Invalid artifact type'

        self._artifact_type = artifact_type

    def _retrieve_artifacts(self, name, limit):
        params = {'q': name, 'type': self._artifact_type,
        'limit': limit}
        return requests.get(TasteDiveService.API_URL, params).
        json()

    @staticmethod
    def _extract_titles(response):
        artifacts = response['Similar']['Results']
        return [artifact['Name'] for artifact in artifacts]

    def similar_titles(self, titles, limit = 5):
        """
        Returns a set of similar titles up to the defined
        limit. Each instance of this class is supposed to work
        only with one artifact type. This type is specified
        during object construction.
        """

        assert 0 < limit <= 50, 'Limit must be in range (0, 50].'
```

```
return {similar_title
        for title in titles
            for similar_title in TasteDiveService._
            extract_titles(
                self._retrieve_artifacts(title, limit))}
```

The following is an example of how to receive similar movies to *Terminator*:

```
>> td.similar_titles(['Terminator'])
{'Aliens', 'Commando', 'First Blood', 'Predator', 'Terminator
Salvation'}
```

The OMDb service (see http://www.omdbapi.com) exposes a Resource API to obtain movie information. We will use it to get ratings for movies, so that we may sort the final result from TasteDive based on this criterion. Listing 8-2 shows the class to communicate with OMDb.

Listing 8-2. omdb_service.py Module to Help Communicate with the OMDb Service

```
import requests

class OMDbService:
    API_URL = 'http://www.omdbapi.com/'

    def __init__(self, api_key):
        self._api_key = api_key

    def retrieve_info(self, title):
        """Returns information about the movie title in JSON
        format."""
        params = {'apikey': self._api_key, 't': title, 'type':
        'movie', 'r': 'json'}
        return requests.get(OMDbService.API_URL, params).json()
```

You will need to obtain an API key to be able to use OMDb. Luckily, you can ask for a free key (with a daily limit of 1000 requests) on the web site (click the API Key tab on the home page). The following is an example of how to get information about the movie *Terminator*:

```
>> omdb = OMDbService('YOUR API KEY HERE')
>> omdb.retrieve_info('Terminator')
{'Title': 'Terminator',
 'Year': '1991',
 'Rated': 'N/A',
 'Released': 'N/A',
 'Runtime': '39 min',
 'Genre': 'Short, Action, Sci-Fi',
 'Director': 'Ben Hernandez',
 'Writer': 'James Cameron (characters), James Cameron
 (concept), Ben Hernandez (screenplay)',
 'Actors': 'Loris Basso, James Callahan, Debbie Medows,
 Michelle Kovach',
 'Plot': 'A cyborg comes from the future, to kill a girl named
 Sarah Lee.',
 'Language': 'English',
 'Country': 'USA',
 'Awards': 'N/A',
 'Poster': 'N/A',
 'Ratings': [{'Source': 'Internet Movie Database', 'Value':
 '6.2/10'}],
 'Metascore': 'N/A',
 'imdbRating': '6.2',
 'imdbVotes': '23',
 'imdbID': 'tt5817168',
 'Type': 'movie',
 'DVD': 'N/A',
```

```
'BoxOffice': 'N/A',
'Production': 'N/A',
'Website': 'N/A',
'Response': 'True'}
```

We are now ready to create our custom movie recommender that will sort offerings based on their ratings. It is possible to use various sources. We will use the Internet Movie Database as a primary source and fall back to imdbRating, as necessary. You can leverage other source, too (like Rotten Tomatoes). Listing 8-3 shows the simple recommender mashup service.

Listing 8-3. simple_movie_recommender.py Module to Offer Similar Movies in Sorted Order

```python
from tastedive_service import TasteDiveService
from omdb_service import OMDbService

class SimpleMovieRecommender:
    PRIMARY_SOURCE = 'Internet Movie Database'

    def __init__(self, omdb_api_key):
        self._omdb = OMDbService(omdb_api_key)
        self._td = TasteDiveService()

    @staticmethod
    def _retrieve_rating(omdb_response):
        for rating in omdb_response['Ratings']:
            if rating['Source'] == SimpleMovieRecommender.
            PRIMARY_SOURCE:
                return float(rating['Value'].split('/')[0])
        return float(omdb_response['imdbRating'])

    def recommendations(self, titles, limit = 5):
        """
```

Return a list of recommended movie titles up to the specified limit.
The items are ordered according to their ratings (from top to bottom).
"""

```
similar_titles = self._td.similar_titles(titles, limit)
ratings = map(
    lambda title:
            SimpleMovieRecommender._retrieve_
            rating(self._omdb.retrieve_info(title)),
    similar_titles)
return list(map(lambda item: item[1],
                sorted(zip(ratings, similar_titles),
                reverse = True)))
```

The following is a session to retrieve ten recommendations for *Terminator* in sorted order:

```
>> smr = SimpleMovieRecommender('YOUR API KEY')
>> smr.recommendations(['Terminator'], 10)
['Aliens',
 'Die Hard',
 'Predator',
 'First Blood',
 'Robocop',
 'Lethal Weapon',
 'Commando',
 'Terminator Salvation',
 'Alien: Resurrection',
 'Robocop 2']
```

Introduction to LensKit for Python

LensKit for Python (LKPY) is an open-source framework for conducting offline recommender experiments. It has a highly modular design that leverages the PyData ecosystem (https://pydata.org). LensKit enables you to quickly fire up a recommender system and play with various algorithms and evaluation strategies. It also integrates with external recommender tools to provide a common control plane including metrics and configuration. In this section we will demonstrate how easy it is to experiment with LensKit for both research and educational purposes. You can install LensKit by issuing conda install -c lenskit lenskit.

Figure 8-3 depicts the package structure of LensKit, while Figure 8-4 shows the partial UML class diagram of the algorithms package. LensKit is very flexible, enabling you to add new algorithms or metrics that should seamlessly integrate with existing artifacts, thanks to the reliance on standardized data structures and frameworks of PyData. This was the major breakthrough compared to the previous Java-based LensKit system that introduced proprietary data formats as well as kept things buried inside the tool (like opinionated evaluation flow, implicit features, and indirect configuration).

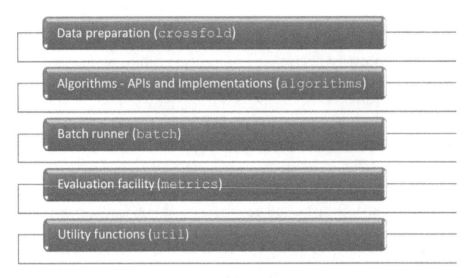

Figure 8-3. *LensKit's package structure (the root package is* lenskit*) with well-defined APIs between them*

The metrics package accepts Pandas Series objects as input, so it may be combined with any external framework. There is a subtle inconsistency of lacking a simple knn wrapper package, since user_knn and item_knn are just different implementations of the nearest neighbor search (something already reflected in the corresponding class names). There are two additional wrapper classes toward the implicit external framework, which are omitted here. The Fallback class is a trivial hybrid that will return the first result from a set of predictors passed as input. This is handy when a more sophisticated algorithm has difficulties computing the output, as a simpler version may provide an alternative answer. I suggest that you read reference [4] for a good overview of various approaches to implement recommendation engines (all examples are realized in RapidMiner, with an extension for recommender systems).

The next example uses the small MovieLens dataset, which you can download from http://files.grouplens.org/datasets/movielens/ ml-latest-small.zip. You should unpack the archive into the data subfolder (relative to this chapter's source code directory). I advise you to read the README.txt file, which is part of the archive, to get acquainted with its content.

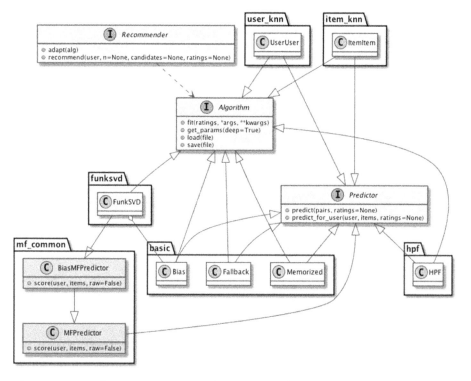

Figure 8-4. *The common API shared by all concrete algorithm classes follows the SciPy style, which enables unified handling of algorithms irrespective of their inner details*

We will demo here the benefits of using LensKit for experimenting with various recommendation engines (see also reference [2]). Just enter the following statements inside an IPython console (make sure you are inside this chapter's source directory); all steps are recorded in the lkpy_demo.py file, so you can easily execute them all by issuing %load lkpy_demo.py. We start with importing the necessary packages:

```
from itertools import tee

import pandas as pd

from lenskit import batch
from lenskit import crossfold as xf
from lenskit.algorithms import funksvd, item_knn, user_knn
from lenskit.metrics import topn
```

Next, we load our dataset and create a Pandas DataFrame object:

```
ratings = pd.read_csv('data/ratings.csv')
ratings.rename({'userId': 'user', 'movieId': 'item'}, axis =
'columns', inplace = True)
print(ratings.head())
   user  item  rating  timestamp
0     1     1     4.0  964982703
1     1     3     4.0  964981247
2     1     6     4.0  964982224
3     1    47     5.0  964983815
4     1    50     5.0  964982931
```

It is important to rename the columns as depicted above. The framework expects users to be under user and items to be under item. Next, we produce a dataset for user-based, five-fold cross-validation:

```
xf_dataset_batch, xf_dataset_test = tee(xf.partition_
users(ratings[['user', 'item', 'rating']]↪
, 5, xf.SampleFrac(0.2)))
truth = pd.concat([test for _, test in xf_dataset_test],
ignore_index = True)
```

The tee function generates two copies of the underlying generator. One is needed for batch evaluation, while the other is needed for creating complete test data. This function is going to be needed later to calculate the ideal DCG for users.

Next we set up a couple of algorithms with different configurations. The MultiEval facility is perfect to sweep over various configuration parameters and produce recommendations. The outcome from each algorithm is saved in a designated subfolder inside a combined Parquet file (you can also make separate files and collect them by calling collect_results). Afterward, we may calculate various metrics to evaluate the achieved level of sophistication.

```
runner = batch.MultiEval('result', False, nprocs = 4)
runner.add_algorithms(
    [item_knn.ItemItem(10), item_knn.ItemItem(20), item_knn.
    ItemItem(30)],
    False,
    ['nnbrs']
)
runner.add_algorithms(
    [user_knn.UserUser(10), user_knn.UserUser(20), user_knn.
    UserUser(30)],
    True,
    ['nnbrs']
)
runner.add_algorithms(
    [funksvd.FunkSVD(40, damping = 0, range = (1, 5)),
     funksvd.FunkSVD(50, damping = 5, range = (1, 5)),
     funksvd.FunkSVD(60, damping = 10, range = (1, 5))],
    False,
```

```
    ['features', 'damping']
)
runner.add_datasets(xf_dataset_batch)
runner.run()
```

The UserUser algorithm is much slower than the other two. This is why you definitely want to run it in parallel. For each algorithm, you can collect pertinent attributes to trace configuration. The next couple of statements load the results and merge them into a nice, unified record structure:

```
runs = pd.read_parquet('result/runs.parquet',
                        columns = ('AlgoClass','RunId','damping'
                        ,'features','nnbrs'))
runs.rename({'AlgoClass': 'Algorithm'}, axis = 'columns',
inplace = True)

def extract_config(x):
    from math import isnan

    damping, features, nnbrs = x
    result = "
    if not isnan(damping):
        result = "damping=%.2f " % damping
    if not isnan(features):
        result += "features=%.2f " % features
    if not isnan(nnbrs):
        result += "nnbrs=%.2f" % nnbrs
    return result.strip()

runs['Configuration'] = runs[['damping','features','nnbrs']].
apply(extract_config, axis = 1)
runs.drop(columns = ['damping','features','nnbrs'], inplace =
True)
```

```
recs = pd.read_parquet('result/recommendations.parquet')
recs = recs.merge(runs, on = 'RunId')
recs.drop(columns = ['RunId'], inplace = True)
print(recs.head(10))
      item     score  user  rank  rating Algorithm Configuration
0    98154  5.273100     3     1     0.0  ItemItem    nnbrs=10.00
1     4429  5.027890     3     2     0.0  ItemItem    nnbrs=10.00
2     1341  5.002217     3     3     0.0  ItemItem    nnbrs=10.00
3   165103  4.991935     3     4     0.0  ItemItem    nnbrs=10.00
4     4634  4.871810     3     5     0.0  ItemItem    nnbrs=10.00
5    98279  4.871810     3     6     0.0  ItemItem    nnbrs=10.00
6     7008  4.869243     3     7     0.0  ItemItem    nnbrs=10.00
7     6530  4.777890     3     8     0.0  ItemItem    nnbrs=10.00
8    32770  4.777890     3     9     0.0  ItemItem    nnbrs=10.00
9     4956  4.777890     3    10     0.0  ItemItem    nnbrs=10.00
```

The extract_config method collects relevant configuration information per algorithm. We will group an algorithm together with its configuration to decide what works best. The next lines compute the nDCG Top-N accuracy metric (see also Exercise 8-1):[2]

```
user_dcg = recs.groupby(['Algorithm', 'Configuration',
'user']).rating.apply(topn.dcg)
user_dcg = user_dcg.reset_index(name='DCG')
ideal_dcg = topn.compute_ideal_dcgs(truth)
user_ndcg = pd.merge(user_dcg, ideal_dcg)
user_ndcg['nDCG'] = user_ndcg.DCG / user_ndcg.ideal_dcg
user_ndcg = user_ndcg.groupby(['Algorithm', 'Configuration']).
nDCG.mean()
```

---

[2]At the time of this writing, the stable version is 0.5.0, which doesn't yet contain the ndcg function (the latest development version already has it). So, you may expect the metrics API to change in the future in incompatible ways.

The next two lines produce the bar plot, as shown in Figure 8-5:

```
%matplotlib inline
user_ndcg.plot.bar()
```

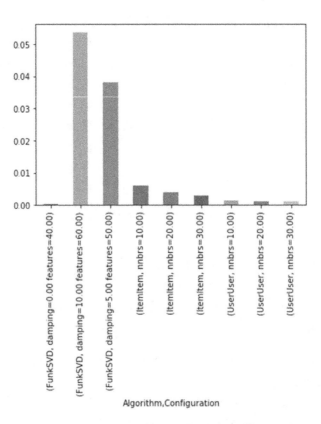

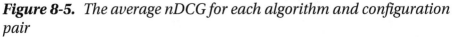

Figure 8-5. *The average nDCG for each algorithm and configuration pair*

LensKit helps you to experiment with a wide range of recommendation algorithms and evaluate them in a standardized fashion. It will be interesting to see whether LensKit will support Predictive Model Markup Language (see http://dmg.org) in the future, as a means of exchanging recommender models. Instead of coding up the pipeline manually, this may come as an XML input formatted according to the PMML schema. PMML serialized models can be managed by GUI tools (like RapidMiner, which can even tune model parameters via generic optimizers, including the one based on a genetic algorithm), which further streamlines the whole process.

EXERCISE 8-1. REPORT PREDICTION ACCURACY

Read about prediction accuracy metrics in LKPY's documentation (`https://lkpy.lenskit.org/en/stable/`). Notice that we have called our evaluator with `batch.MultiEval('result',` **False,** `nprocs = 4)`. Change the second argument to `True` to turn on predictions. You will need to evaluate accuracy using metrics available in the `lenskit.metrics.predict` package.

This will also be a good opportunity to try the `Fallback` algorithm to cope with missing data.

EXERCISE 8-2. IMPLEMENT A NEW METRIC

Among the Top-N accuracy metrics, you will find two classification metrics (a.k.a. decision-support metrics): precision (P) and recall (R). The former reports how many recommended elements are relevant, while the latter reports how many relevant items are considered by a recommender system. Recall is crucial, since this speaks to whether a recommender will really be able to "surprise" you with proper offerings (i.e., avoid missing useful stuff). This may alleviate a known problem called *filter bubble*, where a recommender becomes biased with sparse input provided by a user for specific items. The fact that there are missing ratings for many items doesn't imply that those are worthless for a user.

Usually, we want to balance precision and recall. It is trivial to maximize recall by simply returning everything. F-metrics give us the evenhanded answer. Implement $F_1 = \dfrac{2PR}{P+R}$ in a similar fashion as we have done with nDCG. Rank algorithms based on this new metric.

Summary

In an online world, abundant with information and product offerings, recommender systems may come as a savior. They can select items (news articles, books, movies, music, etc.) matching our interest (assuming a personalized version) and suggest them to us. To diversify the list, most recommenders are hybrids that mix highly customized items and generic items. Moreover, many systems also take into account the current context to narrow down the possibilities. The overall influence of a recommender system may be judged by performing an A/B test and monitoring whether our activities have been altered in a statistically significant way.

Nonetheless, there are two major topics that must be properly handled:

- Privacy and confidentiality of users' preference data, since recommenders may pile up lots of facts based on explicit and implicit input. Users must be informed how this data is handled and why a particular item is recommended (tightly associated with interpretability of machine learning models), and users must be able to manage preference data (for example, you should be able to delete some facts about you from a system). Chapter 9 discusses privacy and confidentiality in depth.

- Negative biases induced by machine-based taste formation. It is a bit disturbing that there are already artists (and consultancy companies to help them) that try to produce content appealing to leading recommender systems. On the other side, to bootstrap the content-based engines, some systems mechanically preprocess and extract features from content (for example, doing signal processing on audio data to discover the rhythm, pitch, and impression of songs).

We definitely wouldn't like to see items at the long tail of any distribution disappear just because they aren't recommender friendly.

References

1. F. Maxwell Harper and Joseph A. Konstan, "The MovieLens Datasets: History and Context," *ACM Transactions on Interactive Intelligent Systems* 5, no. 4, 2015; doi: https://doi.org/10.1145/2827872.

2. Michael D. Ekstrand, "The LKPY Package for Recommender Systems Experiments: Next-Generation Tools and Lessons Learned from the LensKit Project," *Computer Science Faculty Publications and Presentations* 147, Boise State University, presented at the REVEAL 2018 Workshop on Offline Evaluation for Recommender Systems, Oct. 7, 2018; doi: https://doi.org/10.18122/cs_facpubs/147/boisestate; arXiv: https://arxiv.org/abs/1809.03125.

3. Brad Miller, Paul Resnick, Lauren Murphy, Jeffrey Elkner, Peter Wentworth, Allen B. Downey, Chris Meyers, and Dario Mitchell, *Foundations of Python Programming*, Runstone Interactive, https://fopp.umsi.education/runestone/static/fopp/index.html.

4. Vijay Kotu and Bala Deshpande, *Data Science, Concepts and Practice, 2nd Edition*, Morgan Kaufmann Publishers, 2018.

CHAPTER 9

Data Security

Data science is all about data, which inevitably also includes sensitive information about people, organizations, government agencies, and so on. Any confidential information must be handled with utmost care and kept secret from villains. Protecting privacy and squeezing out value from data are opposing forces, somewhat similar to securing a software system while also trying to optimize its performance. As you improve one you diminish the other. As data scientists, we must ensure both that data is properly protected and that our data science product is capable of fending off abuse as well as unintended usage (for example, prevent it from being used to manipulate people by recommending to them specific items or convincing them to act in some particular manner). All protective actions should nicely interplay with the usefulness of a data science product; otherwise, there is no point in developing the product.

Because data security is part of a very broad domain of security engineering, the goal of this chapter is simply to steer your attention to the myriad details you should be aware of. It contains lots of small examples to explain concepts in a pragmatic fashion. It starts with an overview of attack types and tools to combat them. Then, to give you a sense of the types of data privacy regulations that you need to be aware, this chapter elaborates on the European Union's General Data Protection Regulation (GDPR), which affects solutions that process data about EU citizens. Finally, the chapter discusses securing machine learning models to respect privacy.

© Ervin Varga 2019
E. Varga, *Practical Data Science with Python 3*,
https://doi.org/10.1007/978-1-4842-4859-1_9

Checking for Compromise

To get a sense of what privacy actually means, you should visit the site *';--have i been pwned?* (`https://haveibeenpwned.com`) and enter your e-mail address to check if you have an account that has been compromised in a data breach. You may be surprised by the result. Figure 9-1 shows the report I got for one of my e-mail addresses, showing three incidents. The report even contains links to articles explaining in detail how the breaches happened. If your report returns negative results, you should definitely follow the links and read the reports to learn how things work in the "wild."

Breaches you were pwned in

A "breach" is an incident where data has been unintentionally exposed to the public. Using the 1Password password manager helps you ensure all your passwords are strong and unique such that a breach of one service doesn't put your other services at risk.

Adobe: In October 2013, 153 million Adobe accounts were breached with each containing an internal ID, username, email, *encrypted* password and a password hint in plain text. The password cryptography was poorly done and many were quickly resolved back to plain text. The unencrypted hints also disclosed much about the passwords adding further to the risk that hundreds of millions of Adobe customers already faced.

Compromised data: Email addresses, Password hints, Passwords, Usernames

GeekedIn: In August 2016, the technology recruitment site GeekedIn left a MongoDB database exposed and over 8M records were extracted by an unknown third party. The breached data was originally scraped from GitHub in violation of their terms of use and contained information exposed in public profiles, including over 1 million members' email addresses. Full details on the incident (including how impacted members can see their leaked data) are covered in the blog post on 8 million GitHub profiles were leaked from GeekedIn's MongoDB - here's how to see yours.

Compromised data: Email addresses, Geographic locations, Names, Professional skills, Usernames, Years of professional experience

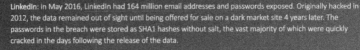

LinkedIn: In May 2016, LinkedIn had 164 million email addresses and passwords exposed. Originally hacked in 2012, the data remained out of sight until being offered for sale on a dark market site 4 years later. The passwords in the breach were stored as SHA1 hashes without salt, the vast majority of which were quickly cracked in the days following the release of the data.

Compromised data: Email addresses, Passwords

Figure 9-1. *The report showing incidents regarding my e-mail*

There are two major problems if your address has been pwned:

- Your private details have possibly leaked out to malevolent persons due to inappropriate security measures.

- Parties you have never contacted know something about you. In my case, I had never visited GeekedIn, although my e-mail was registered in their system; the data about people was collected by inappropriately scraping public GitHub profiles.

This example highlights important mutual responsibilities between entities who collect personal data and users who provide personal data. On one hand, the entities must make best efforts to comply with security standards, and on the other, users must be knowledgeable about dangers and follow general safety advice pertaining to the Internet. For example, users should know to never use the same user account/password combination on multiple web sites (read also reference [1]), never provide sensitive data over an unsecured connection, never click links in suspicious e-mails (how to recognize spam and fishing sites is a topic on its own), and so forth. Computing in the modern world requires adequate security awareness from all participants. This is like being a pedestrian in traffic; you still need to obey basic rules irrespective of whether you have a driver's license or not. Of course, as you aspire to drive a motor vehicle, you must gain enough experience and attain a proper license. Some large vehicles can only be driven by professional drivers. The same applies in data science. As you accumulate more data, you need to ensure that your security skills are at an adequate level, as a potential data breach doesn't only affect you. Many areas of data science are strictly regulated, and companies need to obtain certificates from authorities to handle sensitive data. Various government entities have a variety of different laws and regulations pertaining to data security (we will touch upon the GDPR in the next section as an example). Consult reference [2] to read about an interesting initiative encompassing the whole Web that aims to establish transparency and freedom on the Internet.

Most data science products are remotely accessible via APIs, which allows efficient data exchange scenarios and creation of mashups (introduced in Chapter 8). In all situations, data must be protected both in transit and at rest. Consequently, the external network is usually some secure overlay on the public Web (for example, a secure HTTPS channel, a dedicated virtual private network, etc.). The Internet is a hostile environment, and your product will eventually become a target, especially if it contains precious confidential data.

There are lots of organizations dedicated to sharing best security practices for various technologies, and you should check out at least a few of their web sites. For example, the Open Web Application Security Project (OWASP) is focused on improving the security of software (visit `https://www.owasp.org`) by promoting proven practices and standards. SecTools.org (`https://sectools.org`) compiles many of the best network security tools in one place. (For a good overview of developing secure solutions, check out references [3], [4], and [5].) Other great resources are SecAppDev (`https://secappdev.org`), which is a specialized conference about developing secure products, and the ISO/IEC 27000 family of standards, which prescribes techniques for organizations to keep information assets secure (visit `https://www.iso.org/isoiec-27001-information-security.html`).

More often than not, you will rely on open-source software to implement your data science product (read also the sidebar "OSS Supply Chain Attack"). You have already been exposed to many freely available Python data science frameworks. Even if you theoretically perform everything right, there is a chance that you will introduce a security hole into your product simply by using third-party libraries. Here is a sample of tools that may be useful in this respect:

- Burp Suite (`https://portswigger.net`) helps in finding web security issues.

- Snort (`https://www.snort.org`) is an intrusion prevention system that enables real-time monitoring of networks and taking actions based upon predefined rules.

- Snyk (`https://snyk.io`) automates finding and fixing vulnerabilities in your dependencies. It supports many programming languages, including Python.

- Clair (`https://github.com/coreos/clair`) performs static analysis of vulnerabilities in application containers (supports Docker).

- Docker Bench for Security (`https://github.com/docker/docker-bench-security`) gives valuable advice on how to make your Docker packaged application secure.

- Gitrob (`https://github.com/michenriksen/gitrob`) is a tool to help find potentially sensitive files pushed to public repositories on GitHub. This type of omission is very common, especially when a deployment process is disorganized.

- Vault (`https://www.hashicorp.com/products/vault`) is a tool to solve problems that Gitrob is trying to detect (refer to the previous item).

OSS SUPPLY CHAIN ATTACK

Open-source software (OSS) is a new attack vector where social engineering plays the central role (refer to [6]). A villain starts his journey by making valuable contributions to an open-source project until he gains the confidence of the project sponsors and other contributors. Once he's in possession of proper credentials, he plants vulnerabilities directly into the supply of open-source components. For example, there was a known compromise of PyPI (SSH Decorator), and we are likely to see many more of these types of attack in the future. Due to the huge number of downloads of popular open-source frameworks, the rate at which bogus code spreads can be blistering. Don't forget that if you incorporate open-source components into your own data science product, you are still legally responsible for the security of your product; you are the last link in the chain of responsibility. Therefore, always check the security profile of any open-source component that you consider using.

One way to combat this problem is to resist the temptation to eagerly seek out the latest versions of OSS. Upgrades to versions should be made in a controlled fashion, possibly supported by a centralized enterprise risk management (ERM) tool; a good choice is JFrog Artifactory (see `https://jfrog.com/artifactory`), which has a free community edition. ERM introduces an important level of indirection and may also improve availability by caching artifacts. Developers should only retrieve components approved by the QA department.

Many aspects of security are covered by various standards and tools. Figure 9-2 presents four global categories of security activities that continuously happen throughout the life cycle of your data science product. There are also security models to serve as a template for organizing your security efforts. As an example, you may want to look at the Microsoft Security Development Lifecycle (`https://www.microsoft.com/en-us/sdl`).

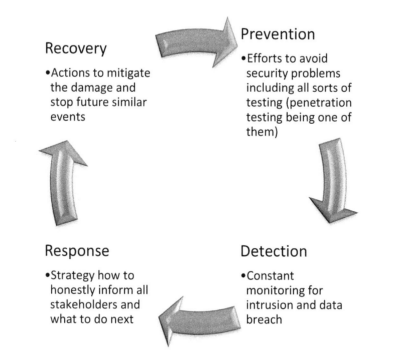

Recovery
- Actions to mitigate the damage and stop future similar events

Prevention
- Efforts to avoid security problems including all sorts of testing (penetration testing being one of them)

Response
- Strategy how to honestly inform all stakeholders and what to do next

Detection
- Constant monitoring for intrusion and data breach

Figure 9-2. *The main phases of handling security events*

It is crucial to have these planned well in advance, when everything works fine. For example, you must verify a backup procedure before the first recovery attempt (assuming that you do have such a procedure established, which is a high priority). Your response to clients must be formulated during a calm period; never attempt to hide the issue! Actually, as you will see later, the GDPR even mandates proper notifications, as specified in Articles 33 and 34 (articles are statements in the GDPR that are treated as obligatory rules).

There are three main ways in which companies may acquire information about you: disclosure, collection, and inference (also watch section 10 of reference [7] and read the sidebar "What Is Privacy?"). Let's use a recommender system as an example (introduced in Chapter 8). If you explicitly rate a movie, then you are voluntarily disclosing information.

Disclosure is the most obvious form of data gathering, since users are fully aware and have control over this process. Collection is a bit more peculiar.[1] Namely, as you click movie descriptions, the system records these events and alters your profile accordingly. Furthermore, *not* clicking a recommended movie is also a useful input, as is the length of time you spend on a page reading about a movie; watching a trailer is another strong indicator that you are attracted to that item. Finally, inference is quite enigmatic. It refers to the capability of machines to discover facts about you based on seemingly innocuous data. The biggest challenge is how to control such inferred information from further abuse. Put another way, if you are the owner of the basic facts, then you should retain ownership over derived values, too. A good proposal that aims to solve this conundrum is given in reference [8].

WHAT IS PRIVACY?

The term *privacy* has different meanings depending on the viewpoint (i.e., it is a subjective notion). A common attribute of privacy is the expectation regarding control. You might perceive unsolicited marketing messages (e-mails, text messages, notifications, etc.) as a privacy violation if you never agreed to receive such stuff and you are given no option to remove your name from the matching address books. From this perspective, the existence or lack of control defines whether you will classify those messages as spam.

There is a subtle tipping point between level of privacy and usefulness/capabilities of data science products. At one extreme, only nonexistent

---

[1]It is interesting to observe that some techniques have gotten reclassified over time due to legal restrictions. In the recent past, web sites were not obligated to warn you about their use of cookies. Nowadays, you must explicitly confirm that you accept the usage of cookies. In other words, collection has turned into disclosure, at least for cookies. Of course, it is a completely different matter whether users are even aware of what those cookies represent. Most of them just click Accept to get rid of the annoying (perhaps intentionally so) pop-up.

(deleted) data is absolutely private. Of course, in that case you can get nothing of value out of a data science service. For example, if you haven't rated any movies or performed any other activity on a movie recommender site, then you shouldn't be surprised that you get purely nonpersonalized content. The other extreme is public data, which obviously contains the highest level of informational content and totally lacks privacy. So, all stakeholders in a data science world must balance privacy and utility. If users can retain control and alter their decisions about the level of privacy, then the basics are properly set.

In her book *Privacy in Context: Technology, Policy, and the Integrity of Social Life* (Stanford University Press, 2009), Professor Helen Nissenbaum suggested that privacy is about *contextual integrity*. In certain contexts, you assume that whatever personal information you divulge in confidence will not be used in unexpected ways or shared without your authorization. For example, if you tell your doctor about your health issues, you assume that the doctor won't share those details with your employer. Therefore, anything tightly associated with a particular behavioral norm is sort of a privacy.

Introduction to the GDPR

It is interesting to peek into this EU law impacting all organizations that handle personal data of EU citizens (including in their Big Data solutions) (also consider watching the video course at [9]). The law applies to both EU-based organizations (regardless of where data processing happens) and non-EU-based organizations handling data about EU citizens (in the rest of the text, by "EU" I mean both EU and European Economic Area [EEA] countries). The GDPR is an incremental evolution of previous data protection laws to accommodate recent technological advances (such as the Internet of Things, Big Data, e-commerce, etc.). Essentially, it is a law that attempts to codify behavioral patterns of organizations to ensure that they respect privacy of people. As a law, it is strictly monitored by local Data Protection Authorities, who enforce accountability for lapses in data security. The idea is to enable interoperable data science solutions with shared expectations about data privacy. Even if you do not work in an EU country, chances are that your organization may be processing data on behalf of an EU-based company. Moreover, it is beneficial to take a look at how privacy may be formulated as a law. The GDPR is not a security standard, so concrete techniques on how to implement it are beyond the scope of this book.

The GDPR is composed of two types of items: 99 *articles* and 173 *recitals.* The former may be approximated with business requirements, while the latter serve as principles that explain the reasons for the articles. I like how the GDPR is presented at `https://gdpr-info.eu` with an additional *key issues* element type (which depicts some common problems in applying the GDPR, with references to articles and recitals).[2] I suggest that you keep Article 4 open while sifting through the GDPR, as it contains definitions of key terms (such as data subject, personal data, controller, processor, etc.).

The major contribution of the GDPR is the mapping of data to rights in regard to control and processing embodied in Recital 1? See `https://gdpr-info.eu/recitals/no-1/`: "The protection of natural persons in relation to the processing of personal data is a fundamental right." The idea is to balance people's fundamental rights and freedoms

---

[2]One key issue is the *right to be forgotten* related to mandatory deletion of data after processing is finished (this is tightly interrelated to data retention policy) or when requested by a data subject. Many ML algorithms cannot remove individual records from a trained model, although anonymization can help to some extent. Microservices-based systems with event-sourced domain models have the same difficulty; one possible solution is to keep event-sourced data encrypted with user-specific keys and delete the selected keys as necessary (without a key, the corresponding record is effectively deleted). At any rate, some flexibility and ingenuity are demanded to deliver a fully compliant GDPR solution, while retaining all the benefits of data science offerings. Deletion of data is a qualified right that must be evaluated by lawyers. Moreover, restoration (recovery) procedures must be modified to ensure deleted data is not accidentally restored.

with personal data processing, including rights of controllers (any organization that decides to do anything with your personal data). For example, if a company possesses extremely sensitive data about you, then a data breach may potentially endanger some of your basic rights, such as your right to preserve your private life and family life (consider the users affected by the 2015 hack of the Ashley Madison web site, a dating site for married people seeking affairs).

Article 5 sets forth the principles relating to processing of personal data. In some sense, this should be your entry point into the GDPR. Article 6 explains in detail what is meant by *lawful processing* (there are six legal reasons for processing personal data). Articles 5 and 6 cover the basic category of data. Article 9 also specifies special categories of personal data together with more stringent processing rules.

Caution   Make sure to always follow a well-documented process for any decision making in relation to personal data processing. Any controller must prove to regulators that proper governance is in place and that all measures to protect privacy are strictly obeyed. This is part of the accountability principle.

Tip   Adhering to the GDPR can help you to streamline your data processing pipeline. Item 1c of Article 5 states the following regarding the processing of personal data: "Personal data shall be adequate, relevant and limited to what is necessary in relation to the purposes for which they are processed ('data minimisation')." So, you are forced to rethink the volume of data at rest and in transit. Reducing the amount of data is definitely beneficial regardless of data security requirements.

Data subjects can send requests to controllers pertaining to their data, as specified in Articles 12–23. These requests can assert the following rights:

- Right of access by the data subject, which encompasses the right to receive a copy of all data as well as meta-data (for example, lineage information regarding where the data came from, or how it is handled).

- Right to rectification, so that the data subject can fix errors in data.

- Right to erasure, so that data may be removed even before the retention period ends. Of course, some erasures cannot be satisfied, for legal reasons.

- Right to restrict processing until erroneous data is fixed or deleted.

- Right to data portability that allows efficient data exchange between controllers (for example, if the data subject chooses to transfer data from one controller to another in an automated fashion) and guarantees the receipt of data from a controller in a standard machine-readable format.

- Right to object to processing of data on the grounds of jeopardizing some of the fundamental personal rights and freedoms.

- Right not to be "subject to" automated decision-making, including profiling.

A controller must respond to any request within one month, free of charge. Definitely, it pays off to automate many tasks, if possible. One particularly dangerous approach is to always let people remotely manage their personal data. The recommendation from Recital 63 related to this is, "Where possible, the controller should be able to provide remote access to a secure system which would provide the data subject with direct access to his or her personal data." Any remote access over the Internet immediately entails a myriad of security issues, and a mistake can be devastating.

There is another set of articles, Articles 24–43, that regulate obligations of controllers and processors. These can be treated like data protection requirements. Article 32 is about security of processing, which contains concrete techniques and measures that controllers and processors need to implement (such as pseudonymization and encryption of data). To better understand the information security measures of the GDPR, consider Recital 83 as a brief summary of Article 32:

> *In order to maintain security and to prevent processing in infringement of this Regulation, the controller or processor should evaluate the risks inherent in the processing and implement measures to mitigate those risks, such as encryption. Those measures should ensure an appropriate level of security, including confidentiality, taking into account the state of the art and the costs of implementation in relation to the risks and the nature of the personal data to be protected. In assessing data security risk, consideration should be given to the risks that are presented by personal data processing, such as accidental or unlawful destruction, loss, alteration, unauthorised disclosure of, or access to, personal data transmitted, stored or otherwise processed which may in particular lead to physical, material or non-material damage.*

—Recital 83, GDPR

The GDPR suggests a *people-oriented, risk-based* security system. This is very important to understand, as any absolutistic approach would definitely fail. Security standards like ISO 27001, NIST, and others may give guidance about appropriate techniques, but they should be applied in the context of the impact on fundamental rights and freedoms of data subjects, rather than organizations, in case of a security breach. Therefore, at the heart of GDPR Article 32 is risk management, which is a broad topic applicable in any software development project. The main difference here is that estimating the severity of risk must include the number of data subjects potentially impacted by an attack.[3] Also, be warned that volume influences both the probability and impact of the risk, so it is imprecise to calculate risk as barely a multiple of probability, impact, and volume. In other words, risk = probability(volume) × impact(volume); both of these quantities are a function of volume and may have nonlinear relationships. For example, the chance that your system is going to be a target of an attack increases as it stores more records. Furthermore, any data breach will have a higher impact, as more data subjects will be hit. As it is hard to express these quantities as real numbers, you may use ordinal numbers instead (take inspiration from T-shirt sizing of agile user stories). For example, a risk with low probability and low impact would have low severity.

Caution   Don't be blinded by numbers. If there is a threat that may jeopardize somebody's life, then the highest risk level applies. I have already mentioned the data breach of the Ashley Madison dating site. At least two people whose data was exposed thereafter committed suicide, so any similar system should treat risks related to data theft as critical.

All risks should be registered and stored in a central place. They should be addressed based on some prioritization strategy, very similar to how user stories are picked from a product backlog. The key difference is that the IT team needs to work closely with the governance/legal department

---

[3]An overall risk is the sum of expected values of random variables, each representing a specific threat. Consequently, you must model threats as part of your risk assessment procedure. There are also tools to support your effort, like the freely available Microsoft Threat Modeling Tool (you can download it from http://bit.ly/ms_threat_modeling_tool).

of an organization rather than let the product owner talk to a customer. The following is a handy checklist to avoid overlooking some risks that are crucial in light of the GDPR:

- Is your capability to remove superfluous data operational? GDPR dictates that all data must be removed after the specified retention period or when requested by a data subject. As a controller you must take care to enforce and supervise data removal at any external processor, especially if it is outside of the EU.

- Do you have areas that you judge are probably noncompliant with the GDPR? This is a special risk type and you should assess risks based upon article numbers. In other words, articles with lower identifiers are heavier than those enumerated later. This means that breaking core principles defined in Article 5 is more severe than not following Article 32.

- Do you have adequate control over processors? Have you arranged what happens after termination of a contract with a processor? You must take care to get back the data or ensure that it is fully erased after completion of the contract.

- Have you planned in enough capacity to promptly respond to requests from data subjects? In some sense, this is an extension of the usual customer support that most organizations already have.

- Is your access control mechanism flexible enough to support all use cases related to processing of data? You may want to apply an attribute-based access control model that comes with a recommended architecture, as shown in Figure 9-3 (see also reference [10]).

- Do you have a documented response plan? You must respond to a data breach not later than 72 hours after becoming aware of such an event. Of course, the event may have happened much earlier, but the 72-hour period doesn't begin until you become aware of the event. Investigation of an issue should continue even after this initial period, to incorporate defensive mechanisms to prevent such occurrences in the future. The regulator will usually help in deciding if and how to contact data subjects.

- Is monitoring of activities of employees in compliance with the GDPR? Employees are also data subjects internal to an organization, so overzealous data collection of their activities might cause privacy problems.

- Is auditing of data processing activities defined and implemented, as described in Article 30?

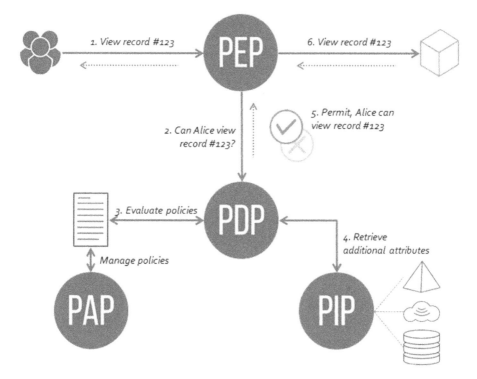

Figure 9-3. *The main components of this architecture are Policy Enforcement Point (PEP), Policy Decision Point (PDP), and Policy Information Point (PIP), as documented by David Brossard at* `http://bit.ly/abac_architecture`

Referring to Figure 9-2, the PEP can be part of the API gateway that intercepts requests and passes the corresponding authorization requests to the PDP. The PDP decides whether to approve or not each request based on predefined policies. These are edited by a separate component, the Policy Authoring Point (PAP). The PDP may gather extra attributes from the PIP, which is a hub that collects attributes from various external sources.

One viable heuristics is to research accepted industry standards and ensure that your practices are commensurate with those of similar enterprises. If you are going to embark into research related to health care, then you may find plenty of peer companies in that same domain. Just ensure to avoid staying behind them. Again, this depends on the context, so take care if you process a special category of data as described in the GDPR.

Article 32 also expands upon availability of your service in relation to its effect on data subjects. This is where domain-driven design and, especially, microservices can help a lot. Having a single database could pose a huge security concern, as it can be a single point of failure (in this case, a target of a data breach). Microservices may choose different technologies to process data, so an attack targeting a specific technology would have a limited scope. Microservices may improve availability as a distributed system, and thus realize the concept of segmentation. Finally, because each microservice is responsible for a dedicated business domain (i.e., has a well-defined bounded context), it is possible to select proper security measures on a case-by-case basis. For example, a health care system may have auxiliary segments that are apart from core services and do not require the same level of protection as patient records. For more information about crafting secure microservices-based systems, watch the video at reference [5].

Another benefit, or positive side effect, of relying on microservices to deliver secure data science products is the high focus on automation and the DevOps paradigm. Centralized logging, monitoring, and distributed tracing (all supported by tools), as well as active execution of robustness and security tests (even in production), aid in complying with Article 32, which has a section about operational awareness and effectiveness of security measures. These are by default part of the DevOps culture that enables having large-scale microservices-based systems in production.

Machine Learning and Security

Attacks on artificial intelligence (AI) systems are a reality. As described in reference [11], researchers have managed to alter just a tiny part of an image used by an AI system in health care to fool the system into believing that some health problem is present, or vice versa. In reference [12], researchers have illustrated how image recognition technology can be fooled in many ways. It is also possible to ferret out personal details from machine learning (ML) models. This imposes another security dimension to be aware of. The major difficulty in assessing an ML-based system's security risk is related to our inability to grasp exactly what is contained inside a trained model. This is associated with interpretability of models, and most models (including those utilized by neural networks) are hard to understand. So, it is impossible to guarantee that a trained model is bullet-proof against some sort of reverse engineering. A similar reasoning applies to ML algorithms (i.e., they are also victims of different attacks).

For a more detailed introduction to the topic of ML security, you may want to read reference [13]. Another superb source of knowledge is the EvadeML project (visit https://evademl.org), which aims to investigate how to make ML methods more robust in the presence of adversaries. In the next few subsections, I will present two types of attack and how they work: a decision-time attack (attack on models) and a training-time attack (attack on algorithms, commonly known as a *poisoning* attack). To make the text lighter and applicable to a broad audience, I will err on the side of accuracy to make matters more comprehensible.

Membership Inference Attack

Let's look at a black-box attack (published in [14]) for figuring out whether some target model was trained using a supervised learning method with a given partial record. This case study is interesting because it leverages an adversarial use of machine learning to train another ML system capable of classifying predictions of the initial system. Apparently, machine learning is used against machine learning to squeeze out information from the target model. The vulnerability of an ML model to this type of attack applies broadly. Nowadays, big cloud providers such as Amazon, Google, Microsoft, and BigML are offering *machine learning as a service (MLaaS)*, and their set of offerings is also susceptible to this sort of attack (actually, they were the primary focus of the attack described in [14]). Thus, having unrealistic expectations regarding the security of ML may put your system at risk. Of course, many weaknesses can be mitigated, but only if you are aware of the issues.

The basic idea is to send a query to the target model and feed its prediction into the adversarial system.[4] The final output would be a signal indicating whether or not data from the query is part of the training dataset. Sending queries and collecting predictions are the only allowable interactions with the target model. (White-box attacks rely on knowing the internal structural details of the target model, so it is an easier problem to solve than the black-box variant.) The premise is that the target model's behavior when it receives input from a training dataset is noticeably different than its behavior when it receives data never seen before by the system. In some sense, this attack belongs to the family of side-channel attacks.[5]

Why should we bother with this? Well, suppose that we have acquired a person's consent to participate in a data science project (consent is one

---

[4]The adversarial ML uses a so-called *attack model* that is purely a naming convention. There is nothing special in this model from the viewpoint of an ML classification algorithm.

[5]An example side-channel attack is a scheme that monitors the fluctuations in power usage of crypto computations to decipher parts of the secret key. The assumption is that nonsecure code leaks out information indicating whether it is processing a binary zero or one at any given moment. Another potent side-channel attack is time based, measuring time differences in data processing to figure out the secret key (this can be leveraged to attack weakly secured web services). You can read more about this in [2].

of the legal bases in GDPR to process personal data). We show her our privacy notice stating that her data will be used only for model training purposes and will be kept secure permanently. Once she agrees to this, any potential data breach from a membership inference attack is automatically a privacy concern. If our data science project stores a special category of data, such as data about health, finance, or a similar sensitive domain, then we should be very vigilant about this potential problem. Be warned that you cannot eschew responsibility by claiming that such leakage wasn't your fault because you are using a commercial cloud service. In the light of the GDPR, as a controller, you are fully liable and accountable for any data processing action carried out by your processor.

Where is the side channel here? In Chapter 7 we talked about the phenomenon of overfitting a model. This happens when we eagerly try to reduce the training prediction error. Consequently, the model picks up tiny details from the training set that are not relevant for generalization. An overfitted model would behave differently depending on whether you provide a record from the training dataset or provide something brand new. The membership inference attack relies on this behavioral difference to categorize the input as either part of or not part of the training dataset (overfitting is not the only possibility, but for the sake of simplicity, assume that it is). Obviously, avoiding overfitting can help prevent membership inference attacks, and regularization is one way to achieve that (see Chapter 7 for details). Hence, improving our ML system to better generalize turns out to be beneficial from the security aspect, too.

Shadow Training

Shadow training is a unique technique whose architecture is depicted in Figure 9-4. It resonates well with the *fundamental theorem of software engineering* coined by David Wheeler: "All problems in computer science can be solved by another level of indirection." In shadow training, shadow models are trained using the same ML platform as was used for the target model. The target system is totally opaque to us, including its private training dataset. The shadow models are trained using synthetic training sets. One simple way to generate a shadow training set is to rely on statistical knowledge about the population from which the training set of the target model was drawn. More specifically, we need to know the marginal distributions of various features and then sample from these distributions. The end result is a statistically valid representative of the private training dataset of the target model. We produce multiple shadow models with different configurations to boost accuracy. The shadow test sets are generated in a similar fashion.

The i-th shadow training set is used to produce the i-th shadow model. This training dataset contains records and their true labels. The output of every shadow model given an input from the matching training dataset is added to the so-called attack training dataset (used to train the final attack model). The test sets are used to create records in the attack training dataset that denote responses of shadow models on unseen data.

The attack model is a simple binary classifier that just indicates whether or not the input belongs to the target's training dataset. Actually, the attack model is a set of models, one model per output class (however, this detail is not that relevant for now). To perform an attack, you pass a known record to the target model and gather its prediction vector. This is a vector of probabilities that represents a distribution of class labels. The highest-scored label is regarded as the predicted class. The attack model receives this prediction vector and the true class label. These were its inputs during the training process (see Figure 9-4). The attack model then outputs YES or NO depending on whether or not it thinks that the given record was part of the target's training dataset.

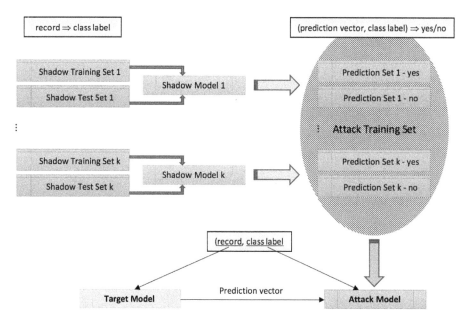

Figure 9-4. *The overall structure of the shadow training process as well as its execution, as described in [14]*

Poisoning Attack

In this type of attack, an attacker modifies the training dataset so that the model gets shifted astray. Usually, there is a cost associated with each modification. There are four main categories of attack:

- **Label modification attack**: The attacker` tries to alter the labels in the dataset. Obviously, this applies only to supervised learning methods. In the case of binary classifiers, this type of action is known as *label flipping*.

- **Poison insertion attack**: The attacker tries to contaminate some limited number of arbitrary feature vectors by adding bogus records.

- **Data modification attack**: The attacker attempts to modify an arbitrary number of records (both features and labels).

- **Boiling frog attack**: An attacker gradually delivers poisoned records in each training iteration. The idea is to keep incremental changes small enough to avoid being detected. Of course, over time the cumulative effect can be significant. The name of the attack comes from a fable whose lesson is that if you put a frog in lukewarm water and slowly increase the water's temperature, then you may cook the frog without the danger that it will jump out of the pot, as it would if you dropped it in boiling water.

Figure 9-5 shows an illustration how a simple linear regression model can be fooled by inserting an outlier. In this case, the obvious remedy would be to remove the outliers, which is a good practice anyway.

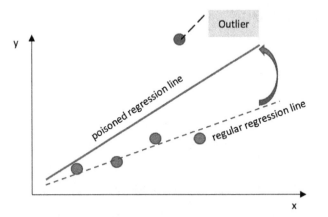

Figure 9-5. *Even a single outlier can move the regression line far away from the planned one*

```
┌─────────────────────────────────────────────────────────────────┐
│            EXERCISE 9-1. STUDY A PRIVACY POLICY                   │
└─────────────────────────────────────────────────────────────────┘
```

Bitly (see `http://bit.ly`) is a popular service to shorten long URLs. It also enables you to track pertinent usage metrics of your links. There is a free plan, so you may start experimenting with the site. Obviously, as it collects information about you (after all, your custom links already reveal a lot about your interests), privacy is a big concern.

Read Bitly's privacy policy document at `https://bitly.com/pages/privacy`. It is written in an approachable style despite being legal text. Besides learning how Bitly handles your profile, you can also get ideas about how to structure a similar artifact for your own data science product.

Summary

Data protection is at the center of data science, since data points are proxies for people and organizations. Cyber-attacks are ubiquitous and will become an even more prevalent problem as smart devices take over bigger responsibilities. This chapter just touched the surface of data security, the goal simply being to steer your attention to this important area. There are plenty of free resources to further expand your knowledge and gain experience in security engineering. One light framework is the Common Sense Security Framework (`https://commonsenseframework.org`), which you could use as a starting point to implement a safe solution. You should also try out Foolbox, a Python toolbox to create adversarial examples for fooling neural networks. It comes with lots of code snippets for trying out various attacks (available at `https://github.com/bethgelab/foolbox`).

You, as a data scientist, must know the limitations of technologies, especially their security aspects. Putting unjustified trust into machines whose decisions may impact humans is very dangerous. We are just at the dawn of the AI era and can expect similar attacks in the future and further revelations about the possibilities and weaknesses of these technologies.

Keeping abreast of the latest AI research in academia is of utmost importance, as most new attacks are described in published research papers.

References

1. Marrian Zhou, "Goodbye Passwords? WebAuthn Is Now an Official Web Standard," CNET, `http://bit.ly/web_authn`, March 4, 2019.

2. Ian Sample, "Tim Berners-Lee Launches Campaign to Save the Web from Abuse," *The Guardian*, `http://bit.ly/magna_carta_web`, Nov. 5, 2018.

3. Ross J. Anderson, *Security Engineering: A Guide to Building Dependable Distributed Systems, Second Edition*, John Wiley & Sons, 2008.

4. Michael Howard, David LeBlanc, and John Viega, *24 Deadly Sins of Software Security: Programming Flaws and How to Fix Them*, McGraw-Hill Professional, 2009.

5. Sam Newman, "Securing Microservices: Protect Sensitive Data in Transit and at Rest," O'Reilly Media, available at `https://learning.oreilly.com/learning-paths/learning-path-building/9781492041481/`, April 2018.

6. Brian Fox, "Open Source Software Is Under Attack; New Event-Stream Hack Is Latest Proof," Sonatype Blog, `http://bit.ly/oss_supply_chain_attack`, Nov. 27, 2018.

7. Ani Adharki, John DeNero, and David Wagner (instructors), "Foundations of Data Science: Prediction and Machine Learning," UC Berkeley

(BerkeleyX) edX Course, `https://www.edx.org/course/foundations-of-data-science-prediction-and-machine-learning-0`, 2018.

8. Rob Matheson, "Putting Data Privacy in the Hands of Users," MIT News, `http://bit.ly/data_control`, Feb. 20, 2019.

9. John Elliott, "GDPR: The Big Picture," Pluralsight Course, `https://www.pluralsight.com/courses/gdpr-big-picture`, 2018.

10. Vincent C. Hu, D. Richard Kuhn, and David F. Ferraiolo, "Access Control for Emerging Distributed Systems," IEEE, *Computer* 51, no. 10 (October 2018): pp. 100–103.

11. Cade Metz and Craig S. Smith, "Warnings of a Dark Side to A.I. in Health Care," *New York Times*, `http://bit.ly/adversarial_ai`, March 21, 2019.

12. Douglas Heaven, "The Best Image-Recognition AIs Are Fooled by Slightly Rotated Images," NewScientist, `http://bit.ly/2UVRXW8`, March 13, 2019.

13. Yevgeniy Vorobeychik and Murat Kantarcioglu, *Adversarial Machine Learning*, Morgan & Claypool Publishers, 2018.

14. Reza Shokri, Marco Stronati, Congzheng Song, and Vitaly Shmatikov, "Membership Inference Attacks Against Machine Learning Models," *2017 IEEE Symposium on Security and Privacy (SP)*, IEEE, `https://www.cs.cornell.edu/~shmat/shmat_oak17.pdf`.

CHAPTER 10

Graph Analysis

A graph is the primary data structure for representing different types of networks (for example, directed, undirected, weighted, signed, and bipartite). Networks are most naturally represented as a set of nodes with links between them. Social networks are one very prominent class of networks. The Internet is the biggest real-world network, easily codified as a graph. Furthermore, road networks and related algorithms to calculate various routes are again based on graphs. This chapter introduces the basic elements of a graph, shows how to manipulate graphs using a powerful Python framework, NetworkX, and exemplifies ways to transform various problems as graph problems. This chapter also unravels pertinent metrics to evaluate properties of graphs and associated networks. Special attention will be given to bipartite graphs and corresponding algorithms like graph projections. We will also cover some methods to generate artificial networks with predefined properties.

The main reason for representing a real-world phenomenon as an abstract network (graph) is to answer questions about the target domain by analyzing the structure of that network. For example, if we represent cities as nodes and roads between them as links, then we can readily use many algorithms to calculate various routes between cities. The criteria could be to find the shortest, fastest, or cheapest routes. In social networks, it is interesting to find the most influential person, which can be mapped to the notion of centrality of a node inside a graph. The ability to associate a problem space with a matching graph structure is already half of the solution.

© Ervin Varga 2019
E. Varga, *Practical Data Science with Python 3*,
https://doi.org/10.1007/978-1-4842-4859-1_10

Usage Matrix As a Graph Problem

The best way to understand the types and elements of a graph is to see things in action using NetworkX (https://networkx.github.io). Figure 10-1 shows the major use cases (UCs) of a generic Internet of Things (IoT) platform. Suppose that you want to optimize the system using an analytical framework called *Usage Matrix*. It provides guidance to pinpoint opportunities for tuning the code base. This technique is applicable in many contexts, including data science products. Table 10-1 depicts the template of a usage matrix. Columns denote individual use cases or functions. Rows represent resources or data nodes. Table 10-2 shows one instance of the usage matrix (the values are made up just to illustrate the technique).

Summing down column j designates the relative volume of accesses against all resources caused by that use case. Summing across row i designates the relative volume of accesses against that resource caused by all use cases. Outliers mark the set of columns/rows that are candidates for optimization; that is, the corresponding use cases and/or resources should be prioritized for performance tuning according to the law of diminishing returns. A particular cell's value is calculated as $a_{ij} = freq_{ij}n_{ij}p_j$, where $freq_{ij}$ denotes the use case j's frequency regarding resource i (how many times use case j visits this resource per unit of time), n_{ij} shows the number of accesses per one visit, and p_j is the relative priority of use case j.

We can reformulate our optimization task as a graph problem. Nodes (vertices) are use cases and resources, while links (edges) are weighted according to the impact of use cases on resources. The "impact" relationships are unidirectional (from use cases toward resources). The graph also shows other relationships, too. For example, there are directional edges denoting specific include and extend UML constructs for use cases. Finally, actors are also associated with use cases using the "interact" bidirectional relationships. Overall, a graph where one node may have parallel links is called a *multigraph*; our graph is a weighted directed multigraph. A graph where all links are bidirectional and have some default weight is an example of an undirected graph.

Now, it is easy to calculate the equivalents of column/row sums. Namely, the sum of links incident to a resource node denotes its row sum (as similar reasoning holds for the use case nodes). We can ask many interesting questions, too. Does actor 1's usage of a system have higher impact on resources than actor 2's? Which actor most intensely utilizes the resources? Which actor interacts with most use cases (directly or indirectly)? All answers can be acquired by simply processing the underlying graph using established graph algorithms.

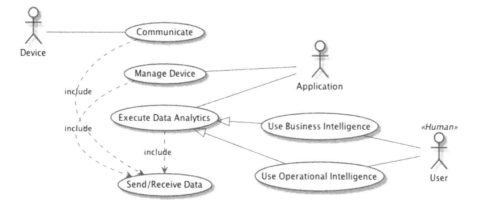

Figure 10-1. *The major use cases and actors in an IoT platform presented as a UML use case diagram (see reference [1] for more information). The usage matrix focuses on use cases and their impact on resources.*

Table 10-1. *Generic Template of a Usage Matrix*

	UC_1	UCj	UCn
R_1			
R_i		a_{ij}	
R_m			

Table 10-2. *Sample Instance of Usage Matrix Template*

	Row Total	Communicate	Manage Dev.	Exec. Data Analytics	Use Bus. Int.	Use Op. Int.
Network	615	120*1*5=600	5*3*1=15	0	0	0
Messaging	700	120*1*5=600	0	20*2*3=100	0	0
Database	2660	0	0	30*10*3=900	40*10*2=800	80*3*4=960
Column Total		1200	15	1000	800	960

The concrete numbers in Table 10-2 are made up and not important. The Send/Receive use case isn't included explicitly in Table 10-2 because it is part of other use cases. Of course, you may structure the table differently. At any rate, this UC will be associated with the other UC via the dedicated relationship type. Note that the UML extend mechanism isn't the same as subclassing in Python and we don't talk here about inheritance. Based on this table, we see that improving the database (possibly using separate technologies for real-time data acquisition and analytics) is most beneficial.

Listing 10-1 shows the code for producing the matching graph using NetworkX. Notice that I have omitted edges with zero weight. The script is quite simple to showcase the ease with which you can construct networks with NetworkX. It is a usual convention to name variables referencing graphs with uppercase letters. You will need to install pydot before executing Listing 10-1 (run conda install pydot from the shell to deploy pydot). Figure 10-2 displays the resulting graph (see also Exercise 10-1).

Listing 10-1. Content of the usage_matrix.py Module

```python
import networkx as nx

G = nx.MultiDiGraph()

# Add all nodes with their role and label. You can immediately
work with labels, but having
# short node identifiers keeps your code uncluttered.
G.add_node(0, role='use-case', label='Communicate')
G.add_node(1, role='use-case', label='Manage Dev.')
G.add_node(2, role='use-case', label='Exec. Data Analytics')
G.add_node(3, role='use-case', label='Use Bus. Int.')
G.add_node(4, role='use-case', label='Use Op. Int.')
G.add_node(5, role='use-case', label='Send/Receive Data')
G.add_node(6, role='resource', label='Network')
G.add_node(7, role='resource', label='Messaging')
G.add_node(8, role='resource', label='Database')
G.add_node(9, role='actor', label='Device')
G.add_node(10, role='actor', label='Application')
G.add_node(11, role='actor', label='User')

# Add edges for the 'impact' relationship.
G.add_edge(0, 6, weight=600, relation='impact')
G.add_edge(0, 7, weight=600, relation='impact')
G.add_edge(1, 6, weight=15, relation='impact')
G.add_edge(2, 7, weight=100, relation='impact')
G.add_edge(2, 8, weight=900, relation='impact')
G.add_edge(3, 8, weight=800, relation='impact')
G.add_edge(4, 8, weight=960, relation='impact')

# Add edges for the 'include' relationship.
G.add_edge(0, 5, relation='include')
G.add_edge(1, 5, relation='include')
G.add_edge(2, 5, relation='include')
```

```
# Add edges for the 'extend' relationship.
G.add_edge(3, 2, relation='extend')
G.add_edge(4, 2, relation='extend')

# Add edges for the 'interact' relationship.
G.add_edge(9, 0, relation='interact')
G.add_edge(0, 9, relation='interact')
G.add_edge(10, 1, relation='interact')
G.add_edge(1, 10, relation='interact')
G.add_edge(10, 2, relation='interact')
G.add_edge(2, 10, relation='interact')
G.add_edge(11, 3, relation='interact')
G.add_edge(3, 11, relation='interact')
G.add_edge(11, 4, relation='interact')
G.add_edge(4, 11, relation='interact')

# Visualize the resulting graph using pydot and Graphviz.
from networkx.drawing.nx_pydot import write_dot

# By default NetworkX returns a deep copy of the source graph.
H = G.copy()

# Set some display properties for specific nodes and extract
labels.
node_labels = {}
for node_id in H.nodes():
    node_labels[node_id] = H.node[node_id]['label']
    role = H.node[node_id]['role']
    if role == 'resource':
        H.node[node_id]['style'] = 'filled'
        H.node[node_id]['fillcolor'] = 'cyan'
        H.node[node_id]['shape'] = 'component'
        H.node[node_id]['fixedsize'] = 'shape'
```

```
    elif role == 'use-case':
        H.node[node_id]['shape'] = 'oval'
    elif role == 'actor':
        H.node[node_id]['style'] = 'rounded'
        H.node[node_id]['shape'] = 'box'
H.node[5]['style'] = 'dashed'
```

nx.relabel_nodes(H, node_labels, copy=False)
pos = nx.nx_pydot.graphviz_layout(H)
nx.draw(H, pos=pos, with_labels=True, font_weight='bold')
write_dot(H, 'usage_matrix.dot')

The script creates the 'usage_matrix.dot' file inside the current directory. It contains instructions for using Graphviz (https://www.graphviz.org) to render the graph. To generate the final graph, you will need to install this tool. Issue the following command from the shell window:

```
dot usage_matrix.dot -Tpng -Gsplines=ortho > usage_matrix.png
```

Caution   I advise you to make a copy of the graph before adding visual effects and relabeling nodes (use the copy parameter to perform the operation in place). If you are not careful with the latter, then you may waste lots of time trying to decipher what is going on with the rest of your code (written to use the previous labels). Keep your original graph tidy, as it should serve as a source of truth.

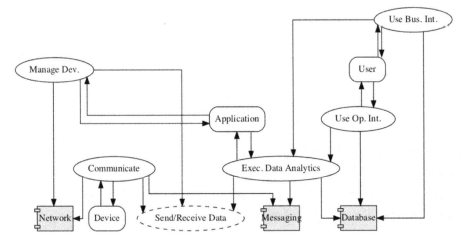

Figure 10-2. *Imagine how much time you would need to spend on this drawing if you weren't able to utilize NetworkX and Graphviz*

For basic drawings, you may rely on the built-in matplotlib engine. However, you shouldn't use it for more complex drawings because, for example, it is quite tedious to control the sizes and shapes of node boxes in matplotlib. Moreover, as you will encounter later, Graphviz allows you to visualize graphs and run algorithms for controlling layouts. Finding out which resource is the most impacted boils down to calculating the degree of resource nodes by using the in_degree method on a graph (keep in mind the directionality of edges). The only caveat is that you must specify the weight argument (in our case, it would point to the custom weight property of associated edges). In an undirected graph, you would simply use the degree method.

Opposite Quality Attributes

It is a known fact that you cannot attain all quality attributes in a product (for a full list, read the ISO/IEC 25010:2011 quality model at bit.ly/iso-25010). Therefore, any optimization effort must find the balance among competing forces. For example, maintainability supports reliability, and vice versa. Performance competes with almost all quality characteristics (which is the reason you should avoid premature optimization and the consequent unnecessary accidental complexity). It is handy to map these relations using a special graph called a *signed graph*, a graph whose edges are denoted with a plus sign or minus sign, depending on whether the matching nodes support or oppose each other. NetworkX allows you to attach arbitrary properties to nodes and edges (we have already seen examples of this, like role, label, relation, etc.). Listing 10-2 shows how you might model the interrelationships between maintainability, reliability, and performance (note that we are dealing here with an undirected graph).

Listing 10-2. Module signed_graph.py Models the Relationships Between Quality Attributes (see Exercise 10-2)

```
import networkx as nx

G = nx.Graph()

G.add_node(0, role='quality-attribute',
label='Maintainability')
G.add_node(1, role='quality-attribute', label='Reliability')
G.add_node(2, role='quality-attribute', label='Performance')

G.add_edge(0, 1, sign='+')
G.add_edge(0, 2, sign='-')
G.add_edge(1, 2, sign='-')
```

Partitioning the Model into a Bipartite Graph

An important category of graphs is bipartite graphs, where nodes can be segregated into two disjunct sets L(eft) and R(ight). Edges are only allowed between nodes from disparate sets; that is, no edge may connect two nodes in set L or R. For example, in a recommender system, a set of users and a set of movies form a bipartite graph for the rated relation. It makes no sense for two users or two movies to rate each other. Therefore, there is only an edge between a user and a movie, if that user has rated that movie. The weight of the edge would be the actual rating.

Let's add a new edge to our first multigraph G between the Communicate use case and the Application actor (see Listing 10-1):

```
G.add_edge(0, 10, relation='interact')
G.add_edge(10, 0, relation='interact')
```

This models the fact that an IoT platform may behave like a virtual device in a federated IoT network. Namely, it can act as an aggregator for a set of subordinate devices and behave as a new device from the perspective of higher-level subsystems. The set of actors and use cases should form a bipartite graph. Listing 10-3 illustrates the procedure to form this type of graph using NetworkX. Here, bipartite.py creates a bipartite graph from our original multigraph by selecting only pertinent nodes and edges. (Note that this code assumes that you have already run usage_matrix.py so that everything is set up.)

Listing 10-3. Module bipartite.py Creating a Bipartite Graph from Our Original Multigraph

```
from networkx.algorithms import bipartite

# Add two new edges as previously described.
G.add_edge(0, 10, relation='interact')
G.add_edge(10, 0, relation='interact')

# Select all nodes and edges from G that participate in
'interact' relation and
# create an undirected graph from them.
H = nx.Graph()
H.add_edges_from((u, v) for u, v, r in G.edges(data='relation')
if r == 'interact')
```

```
# Attach a marker to specify which nodes belong to what group.
for node_id in H.nodes():
    H.node[node_id]['bipartite'] = G.node[node_id]['role'] == 'actor'

nx.relabel_nodes(H, {n: G.node[n]['label'].replace(' ', '\n')
for n in H.nodes()}, copy=False)

print("Validating that H is bipartite: ", bipartite.is_
bipartite(H))

# This is a graph projection operation. Here, we seek to find
out what use cases
# have common actors. The weights represent the commonality
factor.
W = bipartite.weighted_projected_graph(H, [n for n, r in
H.nodes(data='bipartite') if r == 0])

# Draw the graph using matplotlib under the hood.
pos = nx.shell_layout(W)
nx.draw(W, pos=pos, with_labels=True, node_size=800, font_
size=12)
nx.draw_networkx_edge_labels(W, pos=pos,
                                edge_labels={(u, v): d['weight']
                                         for u, v, d in
W.edges(data=True)})
```

NetworkX doesn't use a separate graph type for a bipartite graph. Instead, it uses an algorithm that works on usual graph types. Figure 10-3 shows the weighted projected graph.

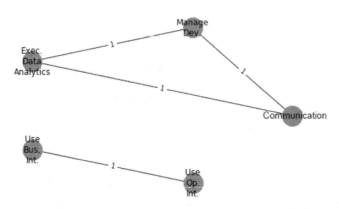

Figure 10-3. *The* `Application` *actor uses all three use cases depicted as a triangle. Therefore, they all have a single common actor. The other two use cases are accessed by the* `User` *actor.*

Scalable Graph Loading

Thus far, we have constructed our graphs purely by code. This is OK for small examples, but in the real world your network is going to be huge. A better approach is to directly load a graph definition from an external source (such as a text file or database). In this section we will examine how to combine Pandas and NetworkX to produce a graph whose structure is kept inside a CSV file (see also the sidebar "Graph Serialization Formats").

GRAPH SERIALIZATION FORMATS

Graphs are usually serialized using the following major formats:

- **Adjacency (incidence) matrix**: This is very similar in structure to our usage matrix, where rows and columns denote nodes. At each intersection of row i and column j there is a value 0 (edge doesn't exist between nodes i and j) or 1 (edge exists between nodes i and j). You can pass this matrix as an initialization argument when creating a new graph. This matrix isn't good to specify additional node/edge properties and is very memory hungry. Graphs are typically sparse, so you will have lots of zeros occupying space.

- **Adjacency list**: A separate row exists for each node. Inside a row there are node identifiers of neighbors. This format preserves space, since only existing edges are listed. An isolated node would have an empty row. You cannot define additional properties with this format. NetworkX has a method `read_adjlist` to read a graph saved as an adjacency list. You can specify the `nodetype` parameter to convert nodes to a designated type. The `create_using` parameter is a pointer to the desired factory (for example, you can force NetworkX to load the graph as a multigraph). Text files may contain comments, and these can be skipped by passing the value for the `comments` parameter.

- **Edge list**: Each row denotes an edge with source and sink node identifiers. You can attach additional items as edge properties (like weight, relation, etc.). It is possible to specify the data types for these properties inside the dedicated `read_edgelist` method. Note that this format alone cannot represent isolated nodes.

- **GraphML**: This is a comprehensive markup language to express all sorts of graphs (visit `http://graphml.graphdrawing.org`). NetworkX has a method `read_graphml` to read graphs saved in this file format.

We will prepare two files: `nodes.csv` to store the nodes (see Listing 10-4), and `edges.edgelist` to describe edges (see Listing 10-5). The first is a CSV file that we will read using Pandas to attach properties to nodes. The other text file is an edge list with the corresponding edge attributes.

Listing 10-6 shows the complete code to load our graph from these files (see the module load_graph.py). NetworkX has a method named from_ pandas_dataframe, but you can't use this to assign node properties (it only supports edge attributes).

Listing 10-4. Content of the nodes.csv File with a Proper Header Row

```
Id,Role,Label
0,use-case,Communicate
1,use-case,Manage Dev.
2,use-case,Exec. Data Analytics
3,use-case,Use Bus. Int.
4,use-case,Use Op. Int.
5,use-case,Send/Receive Data
6,resource,Network
7,resource,Messaging
8,resource,Database
9,actor,Device
10,actor,Application
11,actor,User
```

Listing 10-5. edges.edgelist Stores Edges in the Standard Edge List Format

```
# Add edges for the 'impact' relationship.
0 6 600 impact
0 7 600 impact
1 6 15 impact
2 7 100 impact
2 8 900 impact
3 8 800 impact
4 8 960 impact
```

```
# Add edges for the 'include' relationship. A weight of 1 is
assigned as a placeholder.
0 5 1 include
1 5 1 include
2 5 1 include

# Add edges for the 'extend' relationship.
3 2 1 extend
4 2 1 extend

# Add edges for the 'interact' relationship.
9 0 1 interact
0 9 1 interact
10 1 1 interact
1 10 1 interact
10 2 1 interact
2 10 1 interact
11 3 1 interact
3 11 1 interact
11 4 1 interact
4 11 1 interact
0 10 1 interact
10 0 1 interact
```

Listing 10-6. Complete Code to Load Graph from nodes.csv and edges.edgelist

```
import networkx as nx
import pandas as pd

G = nx.read_edgelist('edges.edgelist',
                     create_using=nx.MultiDiGraph,
                     nodetype=int,
                     data=(('weight', int), ('relation', str)))
```

```
df = pd.read_csv('nodes.csv', index_col=0)
for row in df.itertuples():
    G.node[row.Index]['role'] = row.Role
    G.node[row.Index]['label'] = row.Label

# Make a small report.
print("Nodes: \n", G.nodes(data=True), sep=")
print("-" * 20, "\nEdges: \n", G.edges(data=True), sep=")
```

Running this file produces the following report, assuring us that everything has been properly loaded:

```
Nodes:
[(0, {'role': 'use-case', 'label': 'Communicate'}), (6,
{'role': 'resource', 'label': 'Network'}), (7, {'role':
'resource', 'label': 'Messaging'}), (1, {'role': 'use-case',
'label': 'Manage Dev.'}), (2, {'role': 'use-case', 'label':
'Exec. Data Analytics'}), (8, {'role': 'resource', 'label':
'Database'}), (3, {'role': 'use-case', 'label': 'Use Bus.
Int.'}), (4, {'role': 'use-case', 'label': 'Use Op. Int.'}),
(5, {'role': 'use-case', 'label': 'Send/Receive Data'}), (9,
{'role': 'actor', 'label': 'Device'}), (10, {'role': 'actor',
'label': 'Application'}), (11, {'role': 'actor', 'label':
'User'})]
--------------------

Edges:
[(0, 6, {'weight': 600, 'relation': 'impact'}), (0, 7,
{'weight': 600, 'relation': 'impact'}), (0, 5, {'weight':
1, 'relation': 'include'}), (0, 9, {'weight': 1, 'relation':
'interact'}), (0, 10, {'weight': 1, 'relation': 'interact'}),
(1, 6, {'weight': 15, 'relation': 'impact'}), (1, 5, {'weight':
1, 'relation': 'include'}), (1, 10, {'weight': 1, 'relation':
'interact'}), (2, 7, {'weight': 100, 'relation': 'impact'}),
```

```
(2, 8, {'weight': 900, 'relation': 'impact'}), (2, 5,
{'weight': 1, 'relation': 'include'}), (2, 10, {'weight': 1,
'relation': 'interact'}), (3, 8, {'weight': 800, 'relation':
'impact'}), (3, 2, {'weight': 1, 'relation': 'extend'}), (3,
11, {'weight': 1, 'relation': 'interact'}), (4, 8, {'weight':
960, 'relation': 'impact'}), (4, 2, {'weight': 1, 'relation':
'extend'}), (4, 11, {'weight': 1, 'relation': 'interact'}), (9,
0, {'weight': 1, 'relation': 'interact'}), (10, 1, {'weight':
1, 'relation': 'interact'}), (10, 2, {'weight': 1, 'relation':
'interact'}), (10, 0, {'weight': 1, 'relation': 'interact'}),
(11, 3, {'weight': 1, 'relation': 'interact'}), (11, 4,
{'weight': 1, 'relation': 'interact'})]
```

You can also iterate over nonexistent edges in the graph G by invoking nx.non_edges(G). A similar function that returns the non-neighbors of the node n in the graph G is nx.non_neighbors(G, n).

Social Networks

There are many ways to describe social networks, but we will assume in this section that *social network* means any network of people where both individual characteristics and relationships are important to understand the real world's dynamics. In data science, we want to produce actionable insights from data. The story always begins with a compelling question. To reason about reality and answer that question, we need a model. It turns out that representing people as nodes and their relationships as edges in a graph is a viable model. There are lots of metrics to evaluate various properties (both local and global) of a graph. From these we can infer necessary rules and synthesize knowledge. Moreover, measures of a graph may be used as features in machine learning systems (for example, to predict new link creations based on known facts from a training dataset).

Listing 10-7 shows a simple script to visualize an artificial graph from the atlas of graphs (see reference [2]) and present some distance and centrality measures. For thorough coverage of graph theory, read references [3] and [4]. NetworkX's documentation is an excellent source for additional information about metrics (see the section about algorithms, where you will find subsections about various measures).

Listing 10-7. Module `artificial_network.py`, Which Demonstrates Some Local and Global Graph Metrics

```python
import pandas as pd
import networkx as nx
from networkx.generators.atlas import graph_atlas
import matplotlib.pyplot as plt

G = graph_atlas(681)

# Visualize the graph using the draw_networkx routine.
plt.figure(figsize=(5,5))
nx.draw_networkx(G, node_size=700)
plt.axis('off')
plt.tight_layout();
plt.show()

# Create a data frame to store various centrality measures.
df = pd.DataFrame(index=G.nodes())

df['lcc'] = pd.Series(nx.clustering(G))
df['eccent'] = pd.Series(nx.eccentricity(G))
df['deg-cent'] = pd.Series(nx.degree_centrality(G))
df['clos-cent'] = pd.Series(nx.closeness_centrality(G))
df['btwn-cent'] = pd.Series(nx.betweenness_centrality(G,
normalized=True, endpoints=False))
```

```
print(df)
print('\nAverage clustering coefficient:',
      nx.average_clustering(G))
print('Transitivity:', nx.transitivity(G))
print('Radius:', nx.radius(G))
print('Diameter:', nx.diameter(G))
print('Peripherial nodes:', nx.periphery(G))
print('Central nodes:', nx.center(G))
```

The following is the text output produced by this program, and Figure 10-4 shows the resulting graph:

```
        lcc   eccent   deg-cent   clos-cent   btwn-cent
0   1.000000      3   0.333333        0.50    0.000000
1   1.000000      3   0.333333        0.50    0.000000
2   0.333333      2   0.666667        0.75    0.533333
3   0.500000      2   0.666667        0.75    0.233333
4   0.500000      2   0.666667        0.75    0.233333
5   1.000000      3   0.333333        0.50    0.000000
6   1.000000      3   0.333333        0.50    0.000000

Average clustering coefficient: 0.761904761904762
Transitivity: 0.5454545454545454
Radius: 2
Diameter: 3
Peripherial nodes: [0, 1, 5, 6]
Central nodes: [2, 3, 4]
```

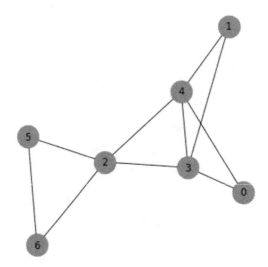

Figure 10-4. *Apparently, you cannot rely on a single metric to decide about some aspect of a graph*

For example, nodes 2, 3, and 4 all have the same closeness centrality. Nonetheless, the betweenness centrality gives precedence to node 2, as it is on more shortest paths in the network. By judiciously selecting and combining different metrics, you may come up with good indicators of a network.

The *local clustering coefficient (LCC)* of a node represents the fraction of pairs of the node's neighbors who are connected. Node 2 has an

$$\text{LCC of } \frac{\#of\ pairs\ of\ node's\ neighbors\ who\ are\ connected}{\#of\ possible\ connected\ pairs} = \frac{2}{\dfrac{d_2(d_2-1)}{2}}$$

$$= \frac{2}{\dfrac{4*3}{2}} = \frac{1}{3}, \text{where } d_2 \text{ is the degree of node 2. LCC embodies people's}$$

tendency to form clusters; in other words, if two persons share a connection, then they tend to become connected, too (a.k.a. *triadic closure*).

Nodes 2, 3, and 4 are considered central because the maximum shortest path (in our case, measured as the number of hops in a path) to any other node is equal to the radius of the network. The radius of a graph is the minimum eccentricity for all nodes. The *eccentricity* of a node is the largest distance between this node and all other nodes. The *diameter* of a graph is the maximum distance between any pair of nodes. Peripheral nodes have eccentricity equal to the diameter. Note that these measures are very susceptible to outliers, in the same way as is the average. By adding a chain of nodes to a network's periphery, we can easily move the central nodes toward those items. This is why social network are better described with different centrality measures.

The *average clustering coefficient* is simply a mean of individual clustering coefficients of nodes. The *transitivity* is another global clustering flag, which is calculated as the ratio of triangles and *open triads* in a network. For example, the set of nodes 2, 5, and 4 forms an open triad (there is a missing link between nodes 5 and 4 for a triangle). Transitivity puts more weight on high-degree nodes, which is why it is lower than the average clustering coefficient (nodes 2, 3, and 4 as central are heavier penalized for not forming triangles).

The centrality measures try to indicate the importance of nodes in a network. *Degree centrality* measures the number of connections a node has to other nodes; influential nodes are heavily connected. The *closeness centrality* says that important nodes are close to other nodes. The *betweenness centrality* defines a node's importance by the ratio of shortest paths in a network including the node. There is a similar notion of betweenness centrality for an edge.

Listing 10-8 introduces kind of a "hello world" for social networks, which is the karate club network (it is so famous that NetworkX already includes it as part of the standard library), as shown in Figure 10-5.

Listing 10-8. karate-network.py Module Exemplifying the Well-Known Karate Club Network (see Exercise 10-3)

```python
import operator

import pandas as pd
import networkx as nx
import matplotlib.pyplot as plt

G = nx.karate_club_graph()

node_colors = ['orange' if props['club'] == 'Officer' else 'blue'
               for _, props in G.nodes(data=True)]
node_sizes = [180 * G.degree(u) for u in G]

plt.figure(figsize=(10, 10))
pos = nx.kamada_kawai_layout(G)
nx.draw_networkx(G, pos,
                 node_size=node_sizes,
                 node_color=node_colors, alpha=0.8,
                 with_labels=False,
                 edge_color='.6')

main_conns = nx.edge_betweenness_centrality(G, normalized=True)
main_conns = sorted(main_conns.items(), key=operator.
itemgetter(1), reverse=True)[:5]
main_conns = tuple(map(operator.itemgetter(0), main_conns))
nx.draw_networkx_edges(G, pos, edgelist=main_conns, edge_
color='green', alpha=0.5, width=6)
nx.draw_networkx_labels(G, pos,
                        labels={0: G.node[0]['club'], 33:
                        G.node[33]['club']},
                        font_size=15, font_color='white')

candidate_edges = ((8, 15), (30, 21), (29, 28), (1, 6))
```

```
nx.draw_networkx_edges(G, pos, edgelist=candidate_edges,
                       edge_color='blue', alpha=0.5, width=2,
                       style='dashed')
nx.draw_networkx_labels(G, pos,
                        labels={u: u for t in candidate_edges
                        for u in t},
                        font_size=13, font_weight='bold',
                        font_color='yellow')

plt.axis('off')
plt.tight_layout();
plt.show()

# Create a data frame to store various centrality measures.
df = pd.DataFrame(index=candidate_edges)

# Add generic and community aware edge features for potential
machine learning classification.
df['pref-att'] = list(map(operator.itemgetter(2),
                          nx.preferential_attachment(G,
                          candidate_edges)))
df['jaccard-c'] = list(map(operator.itemgetter(2),
                          nx.jaccard_coefficient(G, candidate_
                          edges)))
df['aa-idx'] = list(map(operator.itemgetter(2),
                          nx.adamic_adar_index(G, candidate_
                          edges)))
df['ccn'] = list(map(operator.itemgetter(2),
                          nx.cn_soundarajan_hopcroft(G,
                          candidate_edges, 'club')))
df['cra'] = list(map(operator.itemgetter(2),
                          nx.ra_index_soundarajan_hopcroft(G,
                          candidate_edges, 'club')))

print(df)
```

After running this program, you will get the following report:

	pref-att	jaccard-c	aa-idx	ccn	cra
8 15	10	0.400000	0.755386	2	0.000000
30 21	8	0.200000	0.455120	1	0.000000
29 28	12	0.166667	0.352956	2	0.058824
1 6	36	0.083333	0.360674	2	0.062500

The column names designate the following link prediction measures (in the same order from right to left):

- **Preferential Attachment**: Favors nodes with high degree centrality. This is why the score for nodes 1 and 6 is much higher than for the others.

- **Jaccard Coefficient**: The number of common neighbors normalized by the total number of neighbors.

- **Adamic-Adar Index**: Has a logarithm in the denominator that is compared to the Resource Allocation index. For nodes A and B, this index indicates the information flow dispersal level through common neighbors. In a sense, nodes that would like to join tend to have direct communication. If common neighbors have high degree centrality, then such a flow is going to be dissolved.

- **Community Common Neighbors**: The number of common neighbors, with extra points for neighbors from the same community.

- **Community Resource Allocation**: Similar to the Resource Allocation index, but only counts nodes inside the same community. Notice that the candidate edges (9, 15) and (30, 21) have received zero community resources, since they are from different karate clubs.

There are three more real social networks available in NetworkX (as of the time of writing). You may want to experiment with them as we did here. The visualization part can be easily reused.

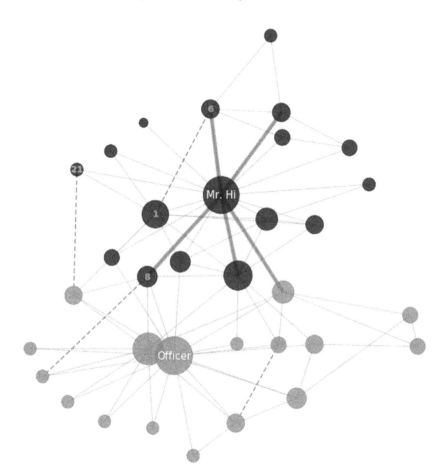

Figure 10-5. *The karate club social network, which clearly emphasizes the split of a club into two parts. The sizes of nodes reflect their degree. The five most important connections are highlighted in green. The dashed lines denote candidate edges. The nodes are colored according to what club they belong to.*

When you study a social network, it is an interesting option to generate realistic network models. These networks aim to mimic pertinent properties of real-world networks under consideration. There are many network generators available inside NetworkX, including the popular preferential model (which obeys the *power law*) and the small-world network (which has a relatively high clustering coefficient and short average path length). The latter tries to mimic the *six degrees of separation* idea. Once you have a model, then you can compare it with observations (samples) from the real world. If the estimated parameters of a model (like the exponent α and constant C in a preferential model) work well to produce networks close to those from the real world, then real-world phenomena may be analyzed by using graph theory and computation. For example, you can train a classifier using network features (like our metrics) to predict probabilities about future connections.

EXERCISE 10-1. BEAUTIFY THE DISPLAYED NETWORK, PART 1

NetworkX coupled with Graphviz gives you tremendous options to customize the visual appearance of a graph. In Figure 10-2 there is no differentiation between relation types or weights. It would be beneficial to present these in a meaningful fashion. Here is a snippet of code to iterate over edges in a directed multigraph G:

```
for i, j, k in G.edges(keys=True):
    print(G.adj[i][j][k])
```

Experiment by setting visual properties of edges (we have set some for nodes). For example, you may color relations differently and set the line widths to the weights of relations. As a bonus exercise, try to attach a custom image for actor nodes. This can be done by mixing in HTML content and referencing an image using the img tag.

EXERCISE 10-2. BEAUTIFY THE DISPLAYED NETWORK, PART 2

Build on the experience you've gained from Exercise 10-1 to visualize a signed graph (like the one we used to model relationships between quality attributes). You may want to display a plus sign or minus sign for each edge as well as color them differently.

You can traverse the edges of an undirected graph G by using the G.edges(data=True) call, which returns all the edges with properties. To display the sign, use the G.edge[node_1][node_2]['sign'] expression, where node_1 and node_2 are node identifiers of this edge.

EXERCISE 10-3. GENERATE AN ERDöS-RÉNYI NETWORK

Read the documentation for the nx.generators.random_graphs.erdos_renyi_graph function to see how to produce this random graph. This model chooses each of the possible edges with probability p, which is a synonym for a binomial graph (since it follows Bernoulli's probability distribution).

The generated network doesn't resemble real-world networks (as a matter of fact, the small-world network model doesn't do a great job in this respect either). Nonetheless, the model is extremely simple, and binomial graphs are used as the basis in graph theory proofs.

Summary

Real-world networks are humungous, so they require more scalable approaches than what has been presented here. One plausible choice is to utilize the Stanford Network Analysis Project (SNAP) framework (visit http://snap.stanford.edu), which has a Python API, too. Storing huge graphs is also an issue. There are specialized database systems for storing graphs and many more graph formats (most of them are documented and supported by NetworkX). One possible selection for a graph database is Neo4j (see `https://neo4j.com/developer/python`).

Graphs are used to judge the robustness of real-world networks (such as the Web, transportation infrastructures, electrical power grids, etc.). Robustness is a quality attribute that indicates how well a network is capable of retaining its core structural properties under failures or attacks (when nodes and/or edges disappear). The basic structural property is connectivity. NetworkX has functions for this purpose (for evaluating both nodes and edges): `nx.node_connectivity`, `nx.minimum_node_cut`, `nx.edge_connectivity`, `nx.minimum_edge_cut`, and `nx.all_simple_paths`. The functions are applicable for both undirected and directed graphs. Graphs are also the basis for all sorts of search problems in artificial intelligence.

References

1. Ervin Varga, Draško Drašković, and Dejan Mijić, *Scalable Architecture for the Internet of Things: An Introduction to Data-Driven Computing Platforms*, O'Reilly, 2018.

2. Ronald C. Read and Robin J. Wilson, *An Atlas of Graphs*, Oxford University Press, 1999.

3. David Easley and Jon Kleinberg, *Networks, Crowds, and Markets: Reasoning About a Highly Connected World*, Cambridge University Press, 2010.

4. Michel Rigo, *Advanced Graph Theory and Combinatorics*, John Wiley & Sons, 2016.

CHAPTER 11

Complexity and Heuristics

As a data scientist, you must solve essential problems in an economical fashion, taking into account all constraints and functional requirements. The biggest mistake is to get emotionally attached to some particular technique and try to find "suitable" problems to which to apply it; this is pseudo-science, at best. A similar issue arises when you attempt to apply some convoluted approach where a simpler and more elegant method exists. I've seen these two recurring themes myriad times. For example, logistic regression (an intuitive technique) often can outperform kernelized support vector machines in classification tasks, and a bit of cleverness in feature engineering may completely obviate the need for an opaque neural network–oriented approach. Start simple and keep it simple is perhaps the best advice for any data scientist. In order to achieve simplicity, you must invest an enormous amount of energy and time, but the end result will be more than rewarding (especially when including long-term benefits). Something that goes along with simplicity is pragmatism—good enough is usually better than perfect in most scenarios. This chapter will exemplify a bias toward simplicity and pragmatism as well as acquaint you with the notion of complexity and heuristics. It contains many small use cases, each highlighting a particular topic pertaining to succinctness and effectiveness of a data product.

© Ervin Varga 2019
E. Varga, *Practical Data Science with Python 3*,
https://doi.org/10.1007/978-1-4842-4859-1_11

Addressing problems while also fully satisfying customers demands two things: a general problem-solving ability and a set of rules and principles that puts searching for a solution into a broader context. The former revolves around the next four steps (see reference [1]), which usually constitute a never-ending cycle if new features are later added to a product:

1. Understanding the problem

2. Devising a plan

3. Carrying out the plan

4. Reviewing and reflecting on the outcome

This strategy is well known and understood by many practitioners. What is usually lesser known is the importance of defining and accepting a framework that will provide systematic guidance in problem-solving endeavors. I prefer the Cynefin framework (shown in Figure 11-1), which is a conceptual model to aid decision-making. There is a tight association between problem solving and decision-making, so forming an ensemble to cope with both topics is vital. Knowing the current domain of the framework helps you in selecting an optimal tactic; you likely would be comfortable choosing a heuristics inside the *Complex* domain, and reluctant to apply such a thing in the *Simple* domain. A *heuristic* is a technique that you use to exchange an absolutely deterministic and accurate solution with something that does the job adequately. Surprisingly, approximate solutions frequently behave as well as fully worked-out methods. Sometimes, the input itself is so noisy that accuracy above some threshold makes no sense.

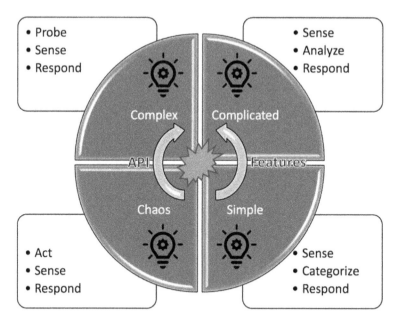

Figure 11-1. *The four quadrants of the Cynefin framework (along with Disorder in the center): Simple, Complicated, Complex, and Chaos*

Without context-awareness, you would encounter disorder and make wrong judgments. There is also a concomitant consequence of applying the Cynefin framework. Namely, to know your current domain, you must think about measurements and metrics. For example, attaining a proper performance characteristic could move you from the *Simple* domain into the *Complicated* or *Complex* domain. However, you will know why you are there and what options you have at your disposal. In other words, you will need to make a deliberate decision about switching a decision-making boundary.

In the center of Figure 11-1 is a special area called Disorder; this represents the situation in which you don't know the right domain yet, and hence must explore further. Each quadrant is linked to a separate problem-solving method instance. The Simple domain calls for a best practice, emphasizing the fact that this is well-charted territory. The

Complicated domain demands a good practice, which may sometimes be an optimal solution. The Complex domain designates an emergent property of a solution, where you may apply various heuristics to reach a suitable resolution. The Chaos domain requires you to take immediate action and only later analyze the cause and effect relationships. Notice that this framework reverses the terms Complicated and Complex compared to the Zen of Python ("Complex is better than complicated," as shown in Figure 3-2 in Chapter 3).

The situation may change over time, and this could cause shifts from one domain into another. New requirements and constraints may demand a different solution, so transitioning happens in counter-clockwise manner. As you gain more knowledge and experience, you could devise simpler solutions, and thus move toward other domains in clockwise direction. Hiding complexity by introducing powerful abstractions in the form of an application programming interface is a viable tactic to simplify things for clients. This means that different stakeholders may look at the whole problem-solution space from different viewpoints. Maybe the best example of this is the generic estimator API of scikit-learn. You may manage many machine learning algorithms in a unified fashion, without worrying about the underlying implementation details, as shown in Figure 11-2.

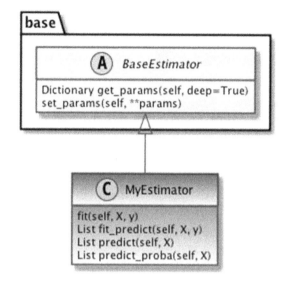

Figure 11-2. *An estimator is an object that handles training and usage of a model*

The base class BaseEstimator implements auxiliary methods for getting and setting parameters of the estimator instance. The `fit`, `fit_predict`, `predict`, and `predict_proba` methods are part of that unifying API, which allows you to treat different estimators in the same way. The `MyEstimator` follows Python's Duck typing convention to deliver the accustomed interface.

From Simple to Complicated

This section presents three major strategies to cope with the Complicated domain: improving the asymptotic running time of an algorithm, applying randomization and sampling, and using parallel/distributed computing. All of them should be potential options in crafting your next data science product. It is very important to start with the simplest solution and gradually enhance it. For this you would need a complete performance test harness, which we are going to showcase here. I have selected the next couple of examples because they are as straightforward as possible while still conveying the main point.

Counting the Occurrences of a Digit

Suppose you need to count how many times a particular digit d appears in numbers inside an interval $[1, n]$, where $n \in \mathbb{N} \land n \leq 10^{50}$. You may start with the naive approach depicted in Listing 11-1. Don't underestimate the utility of such a brute-force program. It may serve well as a test generator (source of truth) for more advanced solutions.

Listing 11-1. Dumb Algorithm That Works Reasonably Well If n Is Under 10^7

```
k, n = tuple(map(int, input().split()))

def count_occurrences_digit_naive(k, n):
    k = str(k)
    count = 0
    for i in range(n + 1):
        count += str(i).count(k)
    return count

print(count_occurrences_digit_naive(k, n))
```

According to the nonfunctional requirements (users should not wait for decades to get the result), we definitely cannot accept this program. It is simply too slow, with roughly linear running time, so we must move beyond the Simple case, as suggested by the Cynefin framework. This is the moment to build the proper infrastructure around performance testing. A similar job is needed to test feeding data into your product, which is part of the data engineering effort. We will use visualization to track time measurements. Listing 11-2 shows the facility to test comparatively different program variants. The functions shown in bold comprise the reusable pieces of this script. The driver code is located inside the conditional block at the end. All you need to do is provide a proper configuration. The attach_model function finds the right constant that goes along with the base model.

Listing 11-2. Performance Test Harness to Track Time Measurements and Visualize Them

```python
import time

import numpy as np
import pandas as pd

def measure(f, num_repetitions=5):
    measurements = np.array([])
    for _ in range(num_repetitions):
        start = time.clock()
        f()
        measurements = np.append(measurements, time.clock() -
        start)
    return measurements.mean()

def execute(config):
    execution_times = {}

    for config_name in config['functions']:
        execution_times[config_name] = np.array([])

    for x in config['span']:
        for config_name in config['functions']:
            execution_times[config_name] = np.append(
                    execution_times[config_name],
                    measure(lambda: config['functions'][config_
                    name](x)))
```

```
    return execution_times

def attach_model(execution_times, config, function_name, model_
name):
    model_vals = np.vectorize(config['models'][model_name])
    (config['span'])
    c = np.mean(execution_times[function_name] / model_vals)
    execution_times[model_name] = c * model_vals

def report(execution_times, x_vals, **plot_kwargs):
    df = pd.DataFrame(execution_times)
    df.index = x_vals
    ax = df.plot.line(
        figsize=(10, 8),
        title='Performance Test Report',
        grid=True,
        **plot_kwargs
    )
    ax.set_xlabel('Span')
    ax.set_ylabel('Time [s]')
    return df

if __name__ == '__main__':
    from count_occurrences_digit_naive import count_
    occurrences_digit_naive

    config = {
        'functions': {
            'naive(k=0)': lambda n: count_occurrences_digit_
            naive(0, n)
        },
        'models': {
            'O(n)': lambda n: n
        },
```

```
    'span': np.geomspace(10**2, 10**7, num=14, dtype=int)
}
execution_times = execute(config)
attach_model(execution_times, config, 'naive(k=0)', 'O(n)')
print(report(execution_times, config['span'], logx=True,
style=['-ro', ':r^']))
```

After executing this code you will get the following report as well as the line plot, as shown in Figure 11-3:

	naive(k=0)	O(n)
100	0.000073	0.000050
242	0.000111	0.000121
587	0.000251	0.000293
1425	0.000608	0.000712
3455	0.001539	0.001725
8376	0.003793	0.004183
20309	0.009700	0.010141
49238	0.023221	0.024587
119377	0.060661	0.059611
289426	0.145676	0.144526
701703	0.361976	0.350397
1701254	0.872631	0.849525
4124626	2.182770	2.059641
10000000	5.334004	4.993522

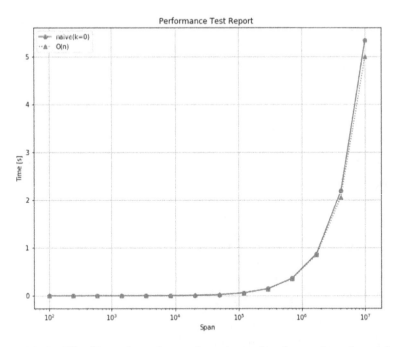

Figure 11-3. The line plot of running times for the naive algorithm and its associated O(n) model. The x axis is on logarithmic scale.

The O(n) model is a slight underestimation, since the body of the loop in Listing 11-1 does depend on the input size. Nevertheless, for very large input sizes (when the running time would be better modeled as $O(n \log n)$), we cannot even wait for the test to finish.

Listing 11-3 shows a much faster O(log n) algorithm that is a result of analyzing the recurring pattern for the number of digits in ranges $[1, 10^i - 1], \forall i \in \mathbb{N} \wedge i \leq 50$. Digits from one to nine produce the same pattern, while zero is a bit different (we need to ignore leading zeros, so the formula is somewhat more challenging). After changing the driver to code to include this faster variant, we get as the line plot shown in Figure 11-4. The difference in execution time is indeed startling. You may modify the test run to cover the full range (you should omit the naive algorithm, as it cannot perform acceptably when $n > 10^7$). It is not hard to generalize the code to handle numbers in arbitrary base (like 16).

406

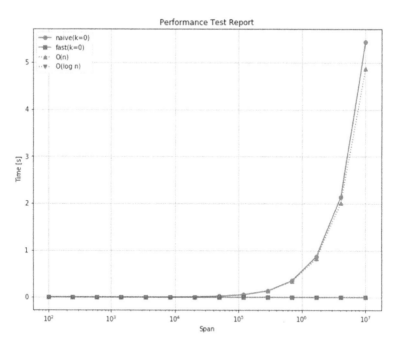

Figure 11-4. *The execution times for the new algorithm are so small that they are all represented as zero at this scale. The O(log n) model is fully aligned with the measured values.*

Listing 11-3. Improved Logarithmic Version That Uses an Additional O(log n) Memory Space.

```
def setup():
    MAX_EXPONENT = 50

    # Holds the number of occurrences of digit k in [0,
    (10**i) - 1], i > 0.
    # If the range is partial (first part of the composite key
    is False), then
    # leading zeros are omitted (this is a special case when k
    == 0).
```

```
    table_of_occurrences = {(False, 0): 0, (False, 1): 1,
                            (True, 0): 0, (True, 1): 1}
    for i in range(2, MAX_EXPONENT + 1):
        table_of_occurrences[(True, i)] = i * 10**(i - 1)
        table_of_occurrences[(False, i)] = \
            10**(i - 1) + 10 * table_of_occurrences[(False,
            i - 1)] - 10
    return table_of_occurrences

def count_occurrences_digit(k, n, table_of_occurrences=setup()):
    digits = str(n)
    num_digits = len(digits)
    count = 0
    is_first_digit = num_digits > 1

    for digit in map(int, digits):
        span = 10**(num_digits - 1)

        count += (digit - 1) * table_of_occurrences[(True, num_
        digits - 1)]
        count += table_of_occurrences[(k != 0 or not is_first_
        digit, num_digits - 1)]

        if digit > k:
            if k > 0 or not is_first_digit:
                count += span
        elif digit == k:
            count += (n % span) + 1

        num_digits -= 1
        is_first_digit = False
    return count
```

```
if __name__ == '__main__':
    k, n = tuple(map(int, input().split()))
    print(count_occurrences_digit(k, n))
```

This case study has demonstrated a situation in which speedup is possible without jeopardizing accuracy. Moreover, the improved code doesn't require any additional infrastructural support. Trying to find a better algorithm should be your top priority (before thinking about Hadoop, Spark, etc.).

Estimating the Edge Betweenness Centrality

In Chapter 10, as part of graph analysis, we examined some centrality metrics for a social network. Among them was the edge betweenness centrality. Calculating the exact value for this metric is CPU intensive. When the network is large and you are interested only in getting a good enough result, then sampling is a viable option. The idea is very simple: You calculate the metric based upon N samples instead over a whole network. This usually gives you nearly identical results, although in a much shorter time. NetworkX has built-in support for this approach. Listing 11-4 shows the code for visually comparing the absolutely accurate outcome with that from sampling. You may want to play around with the code and increase the number of samples. Obviously, as the sample size increases, you will get more accurate figures, but the running time will also increase. You need to experiment to attain the sweet spot.

Listing 11-4. For the sake of terseness, only the first part of the code is repeated, which slightly differs from the original version. Notice the section for evaluating est_main_conns by using 40% of nodes. The karate club network is very small, and sampling is truly beneficial with huge networks. Nonetheless, the principle is the same.

```python
import operator

import networkx as nx
import matplotlib.pyplot as plt

G = nx.karate_club_graph()

node_colors = ['orange' if props['club'] == 'Officer' else 'blue'
               for _, props in G.nodes(data=True)]
node_sizes = [180 * G.degree(u) for u in G]

plt.figure(figsize=(10, 10))
pos = nx.kamada_kawai_layout(G)
nx.draw_networkx(G, pos,
                 node_size=node_sizes,
                 node_color=node_colors, alpha=0.8,
                 with_labels=False,
                 edge_color='.6')

# Calculating the absolute edge betweenness centrality.
main_conns = nx.edge_betweenness_centrality(G, normalized=True)
main_conns = sorted(main_conns.items(), key=operator.
itemgetter(1), reverse=True)[:5]
main_conns = tuple(map(operator.itemgetter(0), main_conns))
nx.draw_networkx_edges(G, pos, edgelist=main_conns, edge_
color='green', alpha=0.5, width=6)
```

```
# Estimating the edge betweenness centrality by sampling 40% of
nodes.
NUM_SAMPLES = int(0.4 * len(G))

est_main_conns = nx.edge_betweenness_centrality
(G, k=NUM_SAMPLES, normalized=True, seed=10)
est_main_conns = sorted(est_main_conns.items(),
key=operator.itemgetter(1), reverse=True)[:5]
est_main_conns = tuple(map(operator.itemgetter(0),
est_main_conns))
nx.draw_networkx_edges(G, pos, edgelist=est_main_conns,
                       edge_color='red', alpha=0.9, width=6,
                       style='dashed')
```

Observe that I've set the seed parameter. This is crucial in order for others to reproduce the results using randomization. Figure 11-5 shows the new graph.

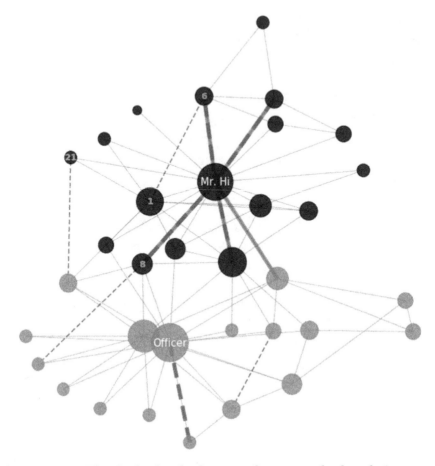

Figure 11-5.   *The dashed red edges are the ones calculated via sampling. Most of them overlap with the original edges, and only one is misconstrued.*

This case study has illustrated how you may exchange accuracy for better performance. This sort of randomization belongs to the class of Monte Carlo methods. The Las Vegas model always produces an accurate result, but in some very rare incidents it may run slower (a good example is quick sort). At any rate, the new code runs faster without reaching out for a computing cluster (see Exercise 11-2).

The Count of Divisible Numbers

Suppose you are given a task to find all natural numbers in the range [a, b] (where $0 < a < b < 10^{18}$) that are divisible by a natural number $c \leq 10^{18}$. Listing 11-5 shows the num_divisible function that solves this problem. It is about as fast as it can be, meaning you cannot considerably speed it up. At this moment you could even stop had there been no additional requirements, like these two:

- There is an enormous number of triples (a, b, c) that you need to address.

- These cannot fit simultaneously in your memory.

These conditions are good telltale signs of the necessity to consider a shift in technology, where applying parallel and distributing computing is indeed justifiable. You can leverage the Dask framework to handle use cases associated with the previous constraints. Listing 11-5 illustrates the core idea behind leveraging a Dask array. The example is tuned down to its essential ingredients so that you can easier comprehend the underlying mechanism (for example, in practice you would work with larger chunks).

Dask delivers constructs (delayed tasks and futures) to parallelize custom applications. Listing 11-5 divides the input array into chunks and processes them as separate blocks. This allows them to be computed in parallel. Dask also handles all the mappings between your array and external data in a transparent fashion. A Dask array is a virtual view into a grid of raw Numpy arrays (the same is true for a Dask dataframe in relation to underlying Pandas dataframes). You work like everything is inside the memory of your machine.

413

Listing 11-5. Dask provides its own version of array and dataframe data structures, whose APIs mimic those found in Numpy and Pandas.

```
import numpy as np
import dask.array as da

def num_divisible(a, b, c):
    r = a % c
    if r == 0:
        start = a
    else:
        start = a + (c - r)

    if start > b:
        return 0
    else:
        return 1 + (b - start) // c

num_divisible_vect = np.vectorize(num_divisible)
x = da.asanyarray([(1, 100, 10), (16789, 445267839, 7), (34,
10**18, 3000), (3, 7, 9)])
x = x.rechunk(chunks=(2, -1))
y = x.map_blocks(lambda block: num_divisible_vect(*block.T),
                 chunks=(-1,),
                 drop_axis=1,
                 dtype='i8')
print(y.compute())
```

By running the preceding program, you will receive the following output, which is an array of four values, one for each input triple:

```
[            10        63607293 333333333333333            0]
```

This case study has demonstrated that parallel and distributed computing is a powerful choice, once you really know that you need it.

From Disorder to Complex

The Complicated domain is characterized by *known unknowns*, while the Complex domain is about *unknown unknowns*. Disorder should be a temporary state while you explore the problem space to understand where it belongs. In this respect, Disorder is the catalyst for exploratory data analysis.

High-dimensional datasets are the most probable reason for initial confusion. Classical visualization doesn't properly operate here, since you don't even know how to represent those dimensions. With high-dimensional data there are lots of duplicated and uninformative features (for example, those that have extremely low or zero variance). These should be dropped from the corpus. Furthermore, many of them will turn out to be unrelated to the target variable, and as such should also be removed as irrelevant. The primary aim is to reduce dimensionality, so that you can see the forest for the trees. The gold standard is the *principal component analysis (PCA)* method, which applies feature extraction to describe your data in terms of orthogonal principal components.

Note   Any approach (SVD, PCA, spectral methods, etc.) that ends up using eigenvalues and eigenvectors inevitably falls into the Complex domain. The main difficulty with these practices is that their output is usually uninterpretable. You can surmise what each eigenvector means, but surely cannot decipher them with certainty.

In this section, we will focus our attention solely on a special visualization technique called t-SNE to glimpse into high-dimensional data; see reference [2] for a nice interactive introduction to this topic as well as how to avoid common pitfalls. In a sense, this is a meta application of the Cynefin framework, since we must consider visualization as Complex/Complicated instead of treating it as Simple.

As a side note, you should also try to peek into the dataset using standard techniques, such as using the `describe` method on the Pandas dataframe. Columns whose variance is near zero are good candidates for removal. Moreover, you can pairwise visualize columns using the `pairplot` function in Seaborn. Highly correlated columns are redundant, and you should avoid them. These actions are all associated with feature selection. Part of this can also be automated. For example, you can use the `sklearn.feature_selection` module that implements feature selection algorithms. For pruning low-variance features, you may use the `VarianceThreshold` class. At any rate, all these methods are OK for low-dimensional datasets (those with less than ten columns/features).

Feature extraction is complementary to feature selection when new features are created. For example, arranging instances into bins (a process called *binning*) is an example of creating new categorical features. We have already encountered this in Chapter 2, where we grouped users by age range.

Exploring the KDD Cup 1999 Data

As stated on the official KDD Cup 1999 Data web site (see `https://kdd.ics.uci.edu/databases/kddcup99/kddcup99.html`):

> This is the data set used for The Third International Knowledge Discovery and Data Mining Tools Competition, which was held in conjunction with KDD-99 The Fifth International Conference on Knowledge Discovery and Data Mining. The competition task was to build a network intrusion detector, a predictive model capable of distinguishing between "bad" connections, called intrusions or attacks, and "good" normal connections. This database contains a standard set of data to be audited, which includes a wide variety of intrusions simulated in a military network environment.

Listing 11-6 shows t-SNE in action for revealing some patterns in the KDD Cup dataset (scikit-learn is packaged with full support for this data). There are 41 features in total. Among them, 34 are numerical, which are only acceptable by t-SNE. Categorical variables should be deselected or encoded. The t-SNE method is very susceptible to even tiny changes. The two auxiliary methods in Listing 11-6 download the column descriptors from the official web site (see also Figure 11-6).

Listing 11-6. Code to Showcase the Power of t-SNE, although you must be patient to try out many different parameter settings (learning rate, random state, perplexity, number of t-SNE components, etc.).

```
from sklearn.datasets import fetch_kddcup99
from sklearn.manifold import TSNE
import pandas as pd
import matplotlib.pyplot as plt
import seaborn as sns

def retrieve_column_desc():
    import requests
    r = requests.get('https://kdd.ics.uci.edu/databases/
    kddcup99/kddcup.names')
    column_desc = {}
    for row in r.text.split('\n')[1:]:
        if row.find(':') > 0:
            col_name, col_type = row[:-1].split(':')
            column_desc[col_name] = col_type.strip()
    return column_desc

def get_numeric_columns(column_desc):
    return [name for name in column_desc if column_desc[name]
    == 'continuous']
```

```
column_desc = retrieve_column_desc()
numeric_columns = get_numeric_columns(column_desc)
print('Number of numeric columns:', len(numeric_columns))
X, _ = fetch_kddcup99(subset='SA', random_state=10, return_X_
y=True)
X = pd.DataFrame(X, columns=column_desc.keys())
X[numeric_columns] = X[numeric_columns].apply(pd.to_numeric)
# We need to work on a small sample to get results in any
reasonable time frame.
X = X.sample(frac=0.05, random_state=10)
m = TSNE(learning_rate=150, random_state=10)
X_tsne = m.fit_transform(X[numeric_columns])
print('First 10 rows of the TSNE reduced dataset:')
print(X_tsne[:10, :])
X['t-sne_1'] = X_tsne[:, 0]
X['t-sne_2'] = X_tsne[:, 1]
sns.set(rc={'figure.figsize': (10, 10)})
sns.scatterplot(x='t-sne_1', y='t-sne_2',
                hue='protocol_type', style='protocol_type',
                data=X[numeric_columns + ['protocol_type',
                't-sne_1', 't-sne_2']])
plt.show()
```

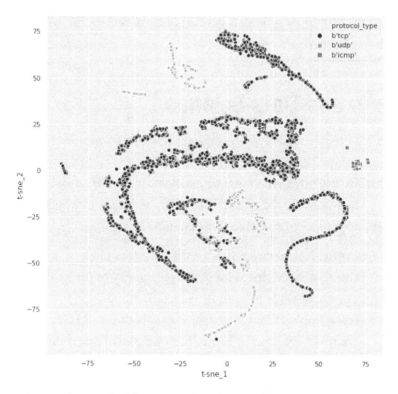

Figure 11-6. *Remarkably, t-SNE has been able to condense 34 features into 2 components, which clearly highlight pertinent patterns in data*

The protocol_type symbolic variable is used to mark various groups (observe the green icmp cluster on the right), and there is a good separation in behavior between them. Once again, t-SNE was able to squeeze out these patterns without looking at the protocol_type feature.

This case study has illuminated the importance of doing exploratory data analysis as part of figuring out into which Cynefin domain your data science problem belongs. Once you know that you need to perform feature extraction and related dimensionality reduction, then you can proceed further by applying PCA or a similar technique. Reducing dimensionality aids your machine learning efforts, since the reduced dataset doesn't require huge storage or processing time, and you reduce the danger of overfitting your model.

Cynefin and Data Science

Some of the previous examples may seem abstract to you, so this section illustrate Cynefin domains through a concrete data science product. Chapter 8 already introduced recommender systems, so we will use them here, too. The following list exemplifies the type of recommendation engine suitable for the first three Cynefin quadrants:

- **Simple**: Nonpersonalized recommenders offer products based on their overall popularity or try to perform various types of association analysis depending on what the current user is doing. For example, if the user is browsing products of a particular type, then the recommender may offer similar products. If a user has put an item into her basket, then the recommender may offer other products commonly bought together with the one inside the basket. This is a typical case when you see on a web page the section "Customers who bought this also bought..."

- **Complicated**: Personalized recommenders utilizing both collaborative and content-based filtering are more intricate and useful than their nonpersonalized counterparts. They try to model customers by developing their profiles. Nowadays, most recommendation engines belong, at least, in this domain.

- **Complex**: Multidimensional and context-aware hybrid recommenders are even more powerful. For example, these recommenders may suggest content based on a user's current mood. This domain seeks out-of-the-box solutions and frequently requires the application of heuristics and approximation algorithms.

EXERCISE 11-1. UNDERSTANDING AN ALGORITHM

For a given closed interval [a, b], where $0 <= a <= b <= 10^8$, the following naive algorithm finds all the numbers with nonrepeating digits and outputs their count:

```
a, b = tuple(map(int, input().split()))

count = 0
for i in range(a, b + 1):
    s = str(i)
    if len(set(s)) == len(s):
        count += 1

print(count)
```

If you provide numbers 0 and 1000, you will get the count of 739. This is a simple but inefficient program with $O((b - a) \log b)$ behavior. The following is a much faster $O((\log b)^3)$ algorithm:

```
import math

a, b = tuple(map(int, input().split()))

def variation_without_repetition(n, k):
    return math.factorial(n) // math.factorial(n - k)
```

```
# Finds how many numbers with nonrepeating digits are present in
# [0, k].
```

```
def count_numbers_with_non_repeating_digits(k):
    if k < 0:
        return 0
    if k == 0:
        return 1

    # We can find most numbers using combinatorics.
    digits = str(k)
    num_digits = len(digits)
    first_digit = int(digits[0])
    span = 10 ** (num_digits - 1)

    s = (first_digit - 1) * variation_without_repetition(9,
    num_digits - 1)

    # We must take care of a lower interval regarding leading zeros.
    s += count_numbers_with_non_repeating_digits(span - 1)

    # We continue our search for the upper part.
    used_digits = {first_digit}
    t = num_digits == 1

    for i in range(1, num_digits):
        first_digit = int(digits[i])
        allowed_digits = set(range(first_digit + 1)) - used_digits
        v = variation_without_repetition(9 - i, num_digits - 1 - i)
        used_digits.add(first_digit)

        if first_digit not in allowed_digits:
            if len(allowed_digits) == 0 and i == 1:
                t = 0
            else:
                t += len(allowed_digits) * v
            break
        else:
            t += (len(allowed_digits) - (i != num_digits - 1)) * v
    return s + t
```

```
print(count_numbers_with_non_repeating_digits(b) - \
      count_numbers_with_non_repeating_digits(a - 1))
```

Try to fully understand how this program works. Apparently, any aggressive performance tuning thwarts the maintainability of your code base. Nonetheless, if you can exponentially speed up your algorithm, that would displace the necessity to utilize expensive and convoluted distributed solutions.

Bonus: Can you generalize the program to handle numbers in an arbitrary base (for example, base 16)? This shouldn't be hard once you have figured out the preceding code. This is the point where the former naive algorithm would immediately break (the search space dramatically increases in bases larger than 10).

EXERCISE 11-2. DATA STREAMING

Suppose you need to produce a histogram of a data stream (i.e., depict the frequency distribution of values in a stream). This sounds like just another counting task. After all, it is simply a matter of using Numpy's `histogram` function for an array of values. The only catch is that the stream is infinite, and values are coming continuously. There is no way to store them in memory. This is another scenario in which you need to change a Cynefin domain from Simple to Complicated.

One very popular algorithm for achieving this is called *Count Sketch*. This is a probabilistic Monte Carlo–style data structure. The idea is to use a 2D array and a set of universal hash functions to count items. Based on the values of counters, you can estimate the frequencies. For more information, read `http://bit.ly/count-sketch`.

Implement a Count-Sketch data structure or find an open-source project and try out this approach. You can simulate a stream using Python's generator function.

EXERCISE 11-3. EXAMINING CHAOS

The Chaos domain requires instantaneous actions. A good example is an anomaly detection system that must react as soon as it detects an abnormal behavior. The KDD Cup 1999 Data is perfect to practice building an automated intrusion detection facility. In this exercise you will do something lighter, but equally instructive.

Hook's law in physics models the force instigated by a spring with some elasticity k when you stretch it by some distance. The formula is $F = -k\Delta x$, where the minus sign denotes an opposite force in relation to the direction of stretching. The spring constant k is unique for each spring.

Suppose that you have collected a set of readings of a force for various Δx displacements. Our model of a spring will be correct as far as the spring is operational. Unfortunately, if you stretch the spring too much, you can damage it. From that moment on you will no longer be able to use the model.

Implement an anomaly detection application for a spring. Your system should prerecord data for regular operational conditions to be able to discern anomalies. You may want to take a look at the `sklearn.neighbors.LocalOutlierFactor` class. When the spring is broken, it will not generate data points near to those from the normal regime. This is the opportunity for your classifier to trigger a warning. You should understand that false positives (invalid alarms) will occur. After all, maybe during normal usage nobody ever tried to push the spring to its limits. Even though nothing bad happened, your system may think that something went wrong. This is the essential property of the Chaos domain. You don't have too much time to think before you act.

Summary

Knowing when a particular approach is proper and admissible is the cornerstone of being a successful data scientist. You must resist the hype and temptation to build your product around a novel technology for the sake of trying things out. Algorithms are at the heart of dealing with Big Data, together with mathematical statistics. This chapter has illustrated how the Cynefin framework provides context for better decision-making and choosing the right solution.

References

1. George Pólya, *How to Solve It: A New Aspect of Mathematical Method*, Expanded Princeton Science Library Edition, Princeton University Press, 2004.

2. Wattenberg, et al., "How to Use t-SNE Effectively," Distill, `https://distill.pub/2016/misread-tsne`, Oct. 3, 2016.

CHAPTER 12

Deep Learning

Deep learning is a field of machine learning that is a true enabler of cutting-edge achievements in the domain of artificial intelligence. The term *deep* implies a complex structure that is designed to handle massive datasets using intensive parallel computations (mostly by leveraging clusters of GPU-equipped machines). The term *learning* in this context means that feature engineering and customization of model parameters are left to the machine. In practice, the combination of these terms in the form of *deep learning* implies multilayered neural networks. Neural networks are heavily used for tasks like image classification, voice recognition/synthetization, time series analysis, and so forth. Neural networks tend to mimic how our brain cells work in tandem in decision-making activities. This chapter introduces you to neural networks and how to build them using PyTorch, which is an open-source Python framework (visit `https://pytorch.org`) that has a familiar API to those accustomed to Numpy. Furthermore, as the last chapter in this book, it exemplifies many stages of the data science life cycle model (data preparation, feature engineering, data visualization, data analysis, and data product deployment). First, though, let's consider the notion of *intelligence* as well as when, how, and why it matters.

© Ervin Varga 2019
E. Varga, *Practical Data Science with Python 3*,
https://doi.org/10.1007/978-1-4842-4859-1_12

Intelligent Machines

The meaning of the term *intelligence* is subjective and has changed over history. In the 18th and 19th centuries, people were amazed by mechanical contrivances and automated steam engines, understandably so; when I visited Tower Bridge in London, I was impressed by the amount of automation present in the coal-driven steam engine system that, until 1974, operated the bascule bridge. When Tower Bridge became operational back in 1894, many Londoners probably thought that the machinery inside Tower Bridge's engine room was "intelligent." Of course, no one today is likely to attach the attribute of intelligent to any sort of steam engine, though. Why? Well, as people's understanding of the mechanics behind a technology becomes commonplace, their associated level of amazement drops accordingly. So, we may establish a correlation between *being intelligent* and our excitement factor regarding some technology. Here are some questions to ponder:

- What makes a car intelligent? Is a self-driving car intelligent? Is a self-parking car also intelligent (albeit not that much as a fully self-driving one)? Is a car with a cruise control system intelligent at all? Where is the borderline?

- Is a common fruit fly intelligent? Is an artificial, self-reproductive fruit fly the *state-of-the-art* of AI in 2019+?

- Are there indirect signs of intelligence? (Take a look at Figure 12-1, discussed a bit later.)

Suppose that you have a "stupid" washing machine without any fuzzy logic or similar smart algorithm built into it. Furthermore, assume that you have stumbled across an "intelligent" washing detergent (such as one marketed at `https://www.skip.co.za/product-format`). Don't contemplate too long about what constitutes intelligence in a washing detergent or how fuzzy logic may help your machine become cleverer.

These aspects aren't important for now. Instead, try to answer the following questions: Would the simple act of pouring an "intelligent" washing detergent into your "stupid" machine make it intelligent? How would you make a judgment? Would you simply evaluate the outcome based on the cleanliness of the washed clothes from the machine? Would you also monitor how your machine operates? Would you disassemble your machine and analyze the parts separately (including the washing detergent), hoping to find intelligence?

Figure 12-1. *You and your team on a mission to find a habitable planet*

Imagine that you and your team are searching nearby solar systems for habitable planets. You approach a good candidate planet, but there is a problem. You have noticed a shovel on the beach. You are afraid to land, not because you are afraid of the shovel, but afraid of what it represents. It is a clear sign of the presence of intelligent beings, who might not welcome you. It doesn't even matter whether the planet is colonized by intelligent living beings or by robots. What matters is the indirect sign of intelligence, which sparks some level of respect. This idea is nicely elaborated in reference [1] and we will further develop it in this chapter.

Estimating the level of intelligence of humans and machines has been a known conundrum for a long time. One way to approach it is embodied in the Turing test, which measures a machine's ability to exhibit intelligent behavior indistinguishable from that of a human, as depicted in Figure 12-2. This is definitely a black-box test, which tries to answer a question purely by observing external properties and outcomes.

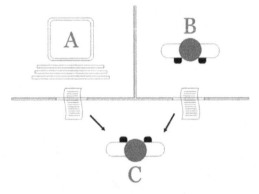

Figure 12-2. *Turing test setup (source:* https://commons. wikimedia.org/wiki/File:Test_de_Turing.jpg). *The interrogator (C) tries to reveal which player (A or B) is a human only by presenting questions and looking at the written answers.*

The Turing test aims to establish the ultimate criteria regarding what constitutes an intelligent machine. Maybe one day we will manage to attain such level of sophistication. Until then we must focus our attention on more pragmatic objectives. Returning to our washing machine example, do we really care who or what cleaned an item of clothing? The real question is, "Can we distinguish whether clothing was washed manually by a human or washed by a washing machine?" If there is no difference, then the washing machine is a useful utensil that makes us more productive and relieve us of mundane tasks. This is the whole point of human–machine interaction. In this sense, intelligence is just a requisite score related to the complexity of tasks that we want to tackle with or without a machine. For more advanced jobs, we need to reach out to more progressed techniques, technologies, and tools. We don't usually care whether the desired level of a machine's intelligence is due to the stupid hardware + super-smart software, ultra-smart hardware + dumb software, or shrewd hardware and software. Sometimes, even an ordinary calculator can make us powerful enough to solve a problem on time with the required level of accuracy.

Note   In the spirit of the previous text, I only contemplate using deep neural networks when simpler, more interpretable solutions are unfeasible. Deep neural networks may seem mystical to external spectators, but internally they are roughly linear algebra and calculus. Nonetheless, the amount of research and human effort that preceded them is staggering (developing the algorithm to train deep neural networks required around 30 years of concerted hard work of many scientists). You may also want to consult references [2] and [3] for more information regarding intelligence and different ways of looking at things.

One of the fascinating and amusing achievements of neural networks is embodied in the 20Q game (for more details, read reference [4]). This product constantly learns from users as they play the game. You may also want to consult reference [5] for examples of when neural networks are a good fit as well as to learn about Keras (I will present PyTorch in this chapter).

Intelligence As Mastery of Symbols

I have pondered a lot about how to best exemplify what is happening inside a neural network. Jumping immediately to nodes, weights, layers, and activation functions seems inappropriate for me. Luckily, I managed to find a suitable problem statement on Topcoder, for a game called AToughGame, to demonstrate how abstractions interact in creating something that appears as intelligent (to follow this discussion, you first must read the specification; click the link for AToughGame at https://www.topcoder.com/blog/how-to-come-up-with-problem-ideas). Intelligence may be treated as a mastery of producing symbols (abstractions) at multiple levels of granularity, as mentioned in reference [1]. Abstract hierarchical structures are accumulated and built upon each other until the final solution may be trivially described in terms of them.

Manual Feature Engineering

Figure 12-3 shows the general structure of the AToughGame problem as a state diagram. The player progresses from one level to another according to the provided probabilities. The main idea is to process the states in pairs. In this manner, the number of states decreases by one after each iteration. This is a typical greedy algorithm with the safe move of aggregating two states. The only real work is then to implement this algorithm with the combine operator. The difficulty is to find out the expected value of a treasure for all possible ways to complete the two levels comprising the pair. The joint probability is simply the product of individual probabilities of levels. All in all, this results in a very fast linear algorithm (visiting each state only once is sufficient).

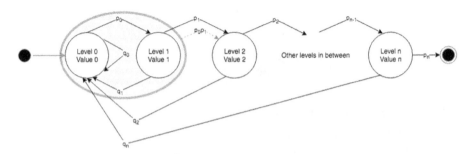

Figure 12-3. *The overall state diagram of the problem*

If the probability of passing some level is p, then the opposite outcome is q = 1 – p. The final state is the winning one. The goal is to calculate the expected amount of treasure collected after completing all levels. The diagram shows the initial pair of states that is going to be aggregated first.

Modeling the Aggregated State

There are two questions to answer about the combined state:

- What is the joint probability of the new state?

- What is the joint value (I will use v to denote a value) of the new state?

We have already answered the first question, as also shown in Figure 12-3. The new value is a sum of the last value and the expected Amount of treasure taking into account all possible ways to leave the two states. The most trivial scenario is to leave the states in succession without dying at either level. The next possibility is dying once at level 0 and afterward finishing both states in succession. The third scenario is to die twice in succession at level 0 and afterward finish both states in succession. This pattern continues indefinitely. We may describe all these scenarios by

$$E(T) = v_1 + v_0 * p_1 * p_0 * \sum_{i=0}^{\infty} q_0^i,$$ where E(T) is the expected amount of treasure.

The last expression already encompasses two powerful abstractions. We should synthesize them now. Don't forget that there are many more possibilities to finish these two levels, which means that manipulating raw probabilities will soon become unwieldy. The first abstraction comes from algebra and provides a closed from solution to the summation. This is called a geometric sum $\sum_{i=0}^{\infty} r^i = \dfrac{1}{1-r}, r < 1.$ In our case, the parameter r will always be a non-negative real number. It is important to name abstractions, and this sum will be denoted as series1(r). Our second abstraction is $\gamma = p_0 * series1(q_0)$, with a meaning of "the probability of leaving level 0." Notice that we can leave this level without dying, or by dying once, twice, and so on.

The next case of leaving the two levels is depicted by the following pseudocode:

```
repeat an arbitrary number of times:
    leave level 0 in any way
    die at level 1
leave level 0 by dying at least once
pass level 1
```

Thanks to the previously introduced abstractions, we may formulate the joint probability for the preceding use case in a succinct fashion as $series1(\gamma * q_1) * q_0 * \gamma * p_1$. Observe how γ is nested inside series1.

The next use case is related to the ability of accumulating wealth by dying multiple times at level 1 without dying at level 0. In other words, this scenario is depicted with the following pseudocode:

```
repeat an arbitrary number of times:
    pass level 0
    die at level 1
pass level 0
pass level 1
```

In this case, after each iteration, the amount of treasure left at level 1 is equal to $v_0 * m$, where m is the number of iterations. To describe this case effectively, we need a new abstraction, which is a derivative of the geometric sum: $\sum_{i=1}^{\infty} i * r^{i-1} = \dfrac{1}{(1-r)^2}, r < 1$. We will name this abstraction as series2(r). Therefore, the joint probability is given by $series2(p_0 * q_1) * p_0 * p_1$.

One final remark is that the last use case can also be associated with the second one. In other words, it is possible to accumulate wealth and then, in the last round, lose everything by leaving level 0 after dying at least once.

Tying All Pieces Together

The whole solution is depicted in Listing 12-1 after refactoring the formulas for each use case. The combine function receives as inputs the raw probabilities and values for two consecutive levels and returns the aggregated probability and expected value of treasure. In between, it builds abstractions from raw data. This is very similar to how neural networks start from raw input nodes, create a hierarchy of abstractions via hidden layers, and finally output the result.

We can get rid of γ as it equals 1. Furthermore, we can simply inline series1 and series2 as well as simplify the expression series1(q1) into 1 / p1. The final result barely resembles the expanded version. Optimization should be left as the last step. These sorts of optimizations also happen inside neural networks. Not all features produced by hidden layers are equally useful nor used in stand-alone manner (many get integrated into higher-level abstractions).

Listing 12-1. AToughGame.py Module That Implements the AToughGame Topcoder Problem

```python
class AToughGame:
    def expectedGain(self, prob, value):
        def combine(level0, level1):
            p0, v0, p1, v1 = level0[0], level0[1], level1[0],
            level1[1]
            q0, q1 = 1 - p0, 1 - p1
            return p0 * p1, v1 + v0 * p1 * (p0 + q0 / p1) *
            (1 - p0 * q1) ** -2
```

```
from functools import reduce
return reduce(combine, zip(map(lambda p: p / 1000,
prob), value))[1]
```

The class name and the sole method's signature are part of the requirements specification (see Exercise 12-1). The internal details, despite all abstractions being manually created, would be unfathomable to someone who hasn't read the preceding description. Therefore, you must take care to accompany your condensed code with a proper design document.

By looking at the preceding code, you might wonder where the intelligence is in these nine lines of code (including the boilerplate stuff and a blank line). Maybe you would marvel at its ingenuity by running it and seeing how it spits out the solution in a couple of nanoseconds. Can you imagine that it has all the necessary knowledge to take into account all possible ways the player could finish the game? This program nicely illustrates the state of affairs in AI before the 1990s. Software solutions were equipped beforehand will all the necessary heuristics and rules.

Now, imagine a software wizard that is capable of deciphering all the necessary abstractions to deal with a problem. This is exactly where neural networks shine. You simply define the architecture of the solution and leave to the network the hard work of producing higher-level symbols. Upper layers reuse abstractions from lower ones. More layers more sophisticated features.

Machine-Based Feature Engineering

We will now build two versions of a neural network to demonstrate automatic feature engineering and how such a network works. The first version will be built from scratch and the second one will use PyTorch. You have already seen linear regression in Chapter 7 (about machine learning), where the output is estimated as $\hat{y} = W * features + b$ (W is the weight of the features and b is the bias term). The features may contain nonlinear components, though. It is possible to associate an activation function with this output; by doing this, you may turn linear regression into logistic regression for classification purposes. An activation function can even be a trivial step function that outputs 1 if $\hat{y} \geq 0$ and 0 otherwise. As a matter of fact, any function that has some threshold to discern positive and negative cases can serve as an activation function.

Figure 12-4 shows a general structure of a neural network with one input layer, one hidden layer, and one output layer. The input layer has as many nodes as there are different features (in this case two) plus an extra constant to represent the bias. The hidden layer is configurable from the viewpoint of number of nodes. The output layer has as many nodes as there are targets (in this case only one).

The shadowed nodes in the hidden layer are composed of an aggregator and an activation function. The node's aggregator calculates the product of the matching input weights and input, while the activation function transforms the value from an aggregator. The final output node may also apply an activation function, when you are doing classification instead of regression. In Figure 12-4, the activation function is the sigmoid function, whose formula is also shown there (its derivative with respect to its input, which is very neat, as you can see).

Every neural network must be trained before use, where training is essentially an iterative and incremental process to find the proper weights. At the beginning, weights are initialized to some small random values. One training cycle (iteration) is composed of two parts: forward pass, which calculates the final output, and backpropagation, which sends back an error through the network to update weights. Cycles are repeated until the network converges to stable weights. Each iteration is traditionally called an *epoch*.

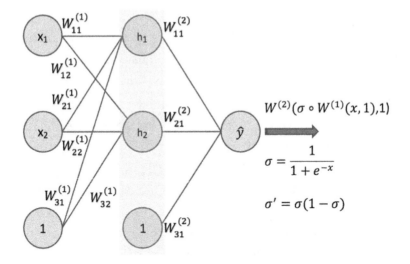

Figure 12-4. *The architecture of a simple neural network for regression task with one hidden layer*

The error is the difference between the true label and the predicted value, represented as $E(W) = y - \hat{y}$. A set of partial derivatives of this error function with respect to weights gives us the desired gradients for updating weights; this is the crux of the gradient descent method. The amount to update the weight $W_{ij}^{(1)}$ is $-\dfrac{\partial E}{\partial W_{ij}^{(1)}} = -\dfrac{\partial E}{\partial \hat{y}} \dfrac{\partial \hat{y}}{\partial h_j} \dfrac{\partial h_j}{\partial W_{ij}^{(1)}}$, which is an application of the chain rule in calculating derivatives. The previous expression would have an additional term had the final output also applied an activation function. At any rate, these partial derivatives are only feasible if the activation function (or functions; you may use different functions in each layer) is continuous and differentiable. Furthermore, to avoid the vanishing gradient issue in deep neural networks (when the product of all partial derivates becomes tiny), the activation function should spread out the output over a larger range. In this respect, the sigmoid function isn't quite good. This is why you will see hyperbolic tangent (tanh) or rectified linear unit (ReLU) in action. The latter is the default choice for hidden layers.

To avoid changing the weights abruptly, there is a hyperparameter called *learning rate*. Every gradient is multiplied by this quantity, so that the process moves cautiously toward a minimum (most often a good local minima).

The power of neural networks comes from the nonlinearity of outputs from hidden nodes. In deep neural networks, as each layer reuses outputs from a previous one, complex features may be created out of raw input. The beauty is that you don't need to worry about how the network will describe the problem in succinct fashion. Of course, the downside is that you will have a hard time interpreting the network's decision-making procedure. There is also an amazing project for producing visual effects from intermediary features created by deep neural networks (visit `https://github.com/google/deepdream`).

Implementation from Scratch

Listing 12-2 provides a full implementation of our simple neural network by only using NumPy arrays. This code is actually my solution for Udacity's bike-sharing project at `http://bit.ly/project-bikes` (I also recommend the excellent free course "Intro to Deep Learning with PyTorch" at Udacity).

Listing 12-2. Simple Neural Network Implementation for Regression Task As Shown in Figure 12-4 (Without Biases)

```
import numpy as np

class NeuralNetwork:
    def __init__(self, input_nodes, hidden_nodes, output_nodes,
    learning_rate):
        self.input_nodes = input_nodes
        self.hidden_nodes = hidden_nodes
        self.output_nodes = output_nodes
```

```
    # Initialize weights to small random values using
    Normal distribution.
    self.weights_input_to_hidden = np.random.normal(
        scale = self.input_nodes ** -0.5,
        size = (self.input_nodes, self.hidden_nodes))
    self.weights_hidden_to_output = np.random.normal(
        scale = self.hidden_nodes ** -0.5,
        size = (self.hidden_nodes, self.output_nodes))

    self.lr = learning_rate
    self.activation_function = lambda x : 1 / (1 +
    np.exp(-x)) # sigmoid

def train(self, features, targets):
    delta_weights_i_h = np.zeros(self.weights_input_to_
    hidden.shape)
    delta_weights_h_o = np.zeros(self.weights_hidden_to_
    output.shape)

    for X, y in zip(features, targets):
        y_hat, hidden_outputs = self.__forward(X)
        delta_weights_i_h, delta_weights_h_o = self.__
        backward(
            y_hat, hidden_outputs,
            X, y,
            delta_weights_i_h, delta_weights_h_o)
    self.__update_weights(delta_weights_i_h, delta_
    weights_h_o)

def run(self, X):
    return self.__forward(X)[0]
```

```
def __forward(self, X):
    hidden_inputs = np.dot(X, self.weights_input_to_hidden)
    hidden_outputs = self.activation_function(hidden_
    inputs)
    final_inputs = np.dot(hidden_outputs, self.weights_
    hidden_to_output)
    y_hat = final_inputs
    return y_hat, hidden_outputs

def __backward(self, y_hat, hidden_outputs, X, y, delta_
weights_i_h, delta_weights_h_o):
    error = y - y_hat
    hidden_error = np.dot(self.weights_hidden_to_output, error)
    output_error_term = error
    hidden_error_term = hidden_error * hidden_outputs *
    (1 - hidden_outputs)
    delta_weights_i_h += np.dot(
        X[:, np.newaxis], hidden_error_term[np.newaxis, :])
    delta_weights_h_o += np.dot(
        hidden_outputs[:, np.newaxis], output_error_
        term[np.newaxis, :])
    return delta_weights_i_h, delta_weights_h_o

def __update_weights(self, delta_weights_i_h, delta_
weights_h_o):
    self.weights_hidden_to_output += self.lr * delta_
    weights_h_o
    self.weights_input_to_hidden += self.lr * delta_
    weights_i_h
```

There are two public methods, `train` and `run`. The private `__update_weights` method doesn't compute the average of the delta weights but assumes that the provided learning rate includes this factor. This is a usual practice, as this parameter is anyhow an arbitrary number that must be tuned for every problem separately.

The weights are initialized to small random values using normal distribution with a standard deviation of $\sqrt{n}$, where n is the number of input nodes into the matching layer. This is known as Xavier initialization, which you may read more about at `http://bit.ly/xavier-init`. This is another way to mitigate the vanishing gradient issue.

Here is a simple recipe to see this network in action:

1. Issue `git clone https://github.com/udacity/deep-learning-v2-pytorch.git` to clone the course repository.

2. Go into the `project-bikesharing` folder and copy there the `simple_network1.py` file from this chapter's source code.

3. Delete the `my_answers.py` file and rename `simple_network1.py` to `my_answers.py`.

4. Open `Predicting_bike_sharing_data.ipynb` in your Jupyter notebook instance and follow the narrative.

The hyperparameters in the `simple_network1.py` file are set as follows:

```
iterations = 1000
learning_rate = 0.005
hidden_nodes = 20
output_nodes = 1
```

All unit tests should pass, and you should get a line plot for the test data as shown in Figure 12-5. The model predicts the data quite well, except for the last week of the year. You can see that in the period of 22 of

December until the end of the year the prediction is higher than the actual data. The network had not been properly trained to recognize this period, when most people take vacation around Christmas. If you look carefully in the notebook, you will see that the test data is not covering a typical period, and this critical period was taken away during training. Furthermore, workingday as a feature was also removed from the data.

Implementation with PyTorch

Listing 12-3 provides a full implementation of our simple neural network, but this time using PyTorch. PyTorch helps you to reduce the amount of code that you need to write, thereby making your product easier to maintain. There is also less chance for you to introduce subtle bugs into your implementation. Most importantly, PyTorch has lots of cool capabilities, like support for deep neural networks, including convolutional, recurrent, gated recurrent, and long-short term memory networks. Since training deep neural networks is computationally quite intensive, PyTorch allows you to utilize GPUs on your machine (if your environment also support the CUDA programming model).

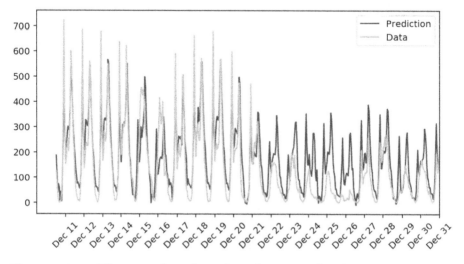

Figure 12-5. *The actual and predicted outputs for the test data*

Listing 12-3. Version of Our Network with PyTorch Using GPUs, If Available The hyperparameters were also altered (look in the `simple_network2.py` module).

```python
from collections import OrderedDict

import torch
from torch import nn

class NeuralNetwork:
    def __init__(self, input_nodes, hidden_nodes, output_nodes,
    learning_rate):
        self.model = nn.Sequential(OrderedDict([
                ('fc', nn.Linear(input_nodes, hidden_nodes)),
                ('sigmoid', nn.Sigmoid()),
                ('output', nn.Linear(hidden_nodes,
                output_nodes))]))

        self.device = torch.device("cuda" if torch.cuda.is_
        available() else "cpu")
        self.model.to(self.device)

        self.criterion = nn.MSELoss()
        self.optimizer = torch.optim.SGD(self.model.
        parameters(), lr = learning_rate)

    def train(self, features, targets):
        features, targets = features.to(self.device),
        targets.to(self.device)

        self.model.train()
        self.optimizer.zero_grad()
        output = self.model(features)
        loss = self.criterion(output, targets)
```

```
        loss.backward()
        self.optimizer.step()

    def run(self, x):
        self.model.eval()
        with torch.no_grad():
            return self.model(torch.tensor(x.values,
            dtype = torch.float) \
                        .to(self.device)).cpu().numpy()
```

The NeuralNetwork class uses composition over inheritance with duck typing and saves the PyTorch network object as an internal attribute model. OrderedDict is useful to name each of the layers of the network. You can easily refer to them later by typing self.model.<name>. The whole code is simply a sequence of declarations instead of low-level implementation details. You can immediately read out the high-level backpropagation algorithm from the body of the train method:

1. Set gradients to zero (the same as we did in Listing 12-2).

2. Make a forward pass through the network.

3. Calculate the error.

4. Backpropagate the error to find the proper deltas for updating the weights.

5. Update the weights.

PyTorch operates with tensors, which are n-dimensional vectors. To run this new version, you will need to alter the code cell inside the Predicting_bike_sharing_data.ipynb notebook for training the network. Don't try to run the unit tests, as they aren't compatible with this code (you may want to tweak them as an additional exercise). Listing 12-4 shows the updated script that uses the PyTorch DataLoader class (the lines shown in bold are additions to the original variant, and some lines have been removed).

Listing 12-4. Updated Script Using PyTorch DataLoader Class

```
import sys
import torch.utils.data as data_utils
from my_answers import iterations, learning_rate, hidden_nodes,
output_nodes

print("Is CUDA available?", "Yes" if torch.cuda.is_available()
else "No")


N_i = train_features.shape[1]
network = NeuralNetwork(N_i, hidden_nodes, output_nodes,
learning_rate)

losses = {'train':[], 'validation':[]}
train = data_utils.TensorDataset(torch.Tensor(np.array
(train_features)),
                                    torch.Tensor(np.array
                                    (train_targets)))
train_loader = data_utils.DataLoader(train, batch_size = 128,
shuffle = True)

for epoch in range(1, iterations + 1):
    for X, y in train_loader:
        network.train(X, y)

    # Printing out the training progress
    train_loss = MSE(network.run(train_features).T,
    train_targets['cnt'].values)
    val_loss = MSE(network.run(val_features).T, val_targets
    ['cnt'].values)
    sys.stdout.write("\rProgress: {:2.1f}".format(100 * epoch /
    iterations) \
```

```
         + "% ... Training loss: " + str(train_
         loss)[:5] \
         + " ... Validation loss: " + str(val_loss)
         [:5])
sys.stdout.flush()

losses['train'].append(train_loss)
losses['validation'].append(val_loss)
```

Figure 12-6 shows how the training and validation losses change over time. After around 200 epochs, there is no significant improvement in either quantity. You should definitely train your network on a beefed-up machine with GPUs, because otherwise it will take a while to finish.

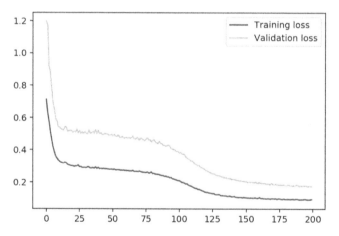

Figure 12-6. *The efficiency of the training process. You should always monitor the validation loss curve. If it starts to rise, then your network is overfitting. If the training loss doesn't drop, then you are underfitting.*

EXERCISE 12-1. CUSTOM PARALLELIZATION

In Chapter 11 we applied Dask to perform operations in parallel over an array. There are situations where this form of concurrency isn't viable. Dask also provides an option to parallelize custom algorithms through the dask. delayed interface.

The expectedGain function returns the expected amount of treasure at the end of a game. Suppose that you want to simply return the total probability of passing all levels in succession. This is currently returned as the first element in the final tuple. Create a new method totalProbability that calculates just this quantity in parallel.

Take a look at the example code about tree summation (reduction) at https://examples.dask.org/delayed.html. Instead of using the add operation from the Dask tutorial, you would use multiplication to merge nodes. For this simple case of tree reduction, you don't need a custom procedure, but the aim is to try out the dask.delayed interface and monitor in the Dask dashboard how computations are carried out.

EXERCISE 12-2. DEPLOYMENT INTO PRODUCTION

PyTorch (starting from version 1.0) offers the capability to convert your Python model into an intermediary format that could be utilized from a C++ environment. In this manner, you can develop your solution in Python and later deploy it as a C++ application. This approach addresses the performance requirements associated with production setup.

Consult the tutorial about loading your PyTorch model in C++ at https:// pytorch.org/tutorials/advanced/cpp_export.html. In our case, because the forward implementation is unified (there is no conditional logic based on input), you can transform the PyTorch model to Torch Script via tracing.

Summary

Using PyTorch, or some other framework for neural networks, is essential to cope with the inherent complexities of deep neural networks. There are lots of additional options to optimize the training process, which are readily available in PyTorch: dropout, batch normalization, various advanced optimizers, different activation functions, and so on. PyTorch also allows you to persist your network into external storage for later use. You might want to save your model each time you manage to reduce the validation loss. At the end, you can select the best-performing variant.

PyTorch is bundled with two major extensions: pytorchvision, which is useful for image processing, and pytorchtext, which is useful for handling text (such as doing sentiment analysis). You can also reuse publicly available trained models to realize the concept of transfer learning. For example, there are models trained on images from ImageNet (`http://www.image-net.org`) with cool features for image classification.

There is an interesting initiative called PyTorch Hub (`https://pytorch.org/hub`) for efficiently sharing pretrained models, thus realizing the vision of transfer learning. Neural networks are also a very popular option at the IoT edge. For this you need an ultra-light prepacked engine. One example is Intel's Neural Compute Stick (`http://bit.ly/neural-stick`). All in all, there are very innovative and interesting approaches for each use case and domain.

References

1. Douglas R. Hofstadter, *Gödel, Escher, Bach: An Eternal Golden Braid*, Anniversary Edition, Basic Books, 1999.

2. Charles Petzold, *Code: The Hidden Language of Computer Hardware and Software*, Microsoft Press, 2000.

3. Garry Kasparov, "Don't Fear Intelligent Machines. Work with Them," TED2017, `https://www.ted.com/talks/garry_kasparov_don_t_fear_intelligent_machines_work_with_them`, April 2017.

4. Karen Schrock, "Twenty Questions, Ten Million Synapses," Scienceline, `https://scienceline.org/2006/07/tech-schrock-20q`, July 28, 2006.

5. Jojo Moolayil, *Learn Keras for Deep Neural Networks: A Fast-Track Approach to Modern Deep Learning with Python*, Apress, 2018.

Index

A

Active learning, 258

Adamic-adar index, 392

Algorithm, 421

Alpha Vantage's API, 288

Anaconda Distribution, 18

    managing packages, 20–22

    Miniconda, 18

    reproducing environments, 23, 24

    Spyder launch, 18

Apache Arrow, 61

Apache Parquet, 60

Application programming interfaces (APIs), 73

    CDCs, 108

    fibonacci numbers, 109, 110

    generic sequencer function, 111

    law of diminishing returns, 108

    roles, 107

AToughGame problem, 432

attach_model function, 403

Augmented ball descend

    problem specification, 159, 160

    version 1.1

        boundaries and movement, 163

        path finding engine, 163, 164, 166–169

        retrospectives, 171, 172

        satellite image dataset, 160–162

    version 1.2

        enhancing input subsystem, 173–177

        enhancing output subsystem, 178–180

        retrospective, 181–184

    version 1.3

        approaches, 184

        baseline, 186, 188

        nonrecursive simple path finder, 189, 191

        ParallelSimplePathFinder class, 197, 198

        performance optimization, 192–195, 197

        retrospective, 200

        reuse-based software engineering characteristics, 185, 186

        test terrain function, 187

Average clustering coefficient, 387, 389

© Ervin Varga 2019

E. Varga, *Practical Data Science with Python 3*,

https://doi.org/10.1007/978-1-4842-4859-1

B

Betweenness centrality, 389,
409–412
Ball Descend project
data model
code, 138
matrix function, 140
NumPy, 138
print statement, 139
problem specification,
136, 137
simulator, 145, 146
test automation, 146–148
Bernoulli's probability
distribution, 395
Bias-variance trade-off, 275
Big Data
four Vs, 9–11
LHC, 12
MOOC, 10, 11
old *vs.* modern data science
projects, 11
Bipartite graphs
application actor, 380
bipartite.py module, 378, 379
IoT platform, 378
rated relation, 377
Boiling frog attack, 364
Boolean indexing technique, 44
Brown Cow model, 3, 4
Brute-force program, 402

C

Closeness centrality, 389
Chaos domain, 400, 424
Click-through rate (CTR), 33, 46,
51, 57, 59
Closest pair algorithm
API-centric and object-oriented
development, 219–223
Euclidian distance, 218
interactive information
radiators
dashboards, 241–243
DSL, 244, 246–249
tabular presentation data, 227
version 1.0
Brute-Force
implementation, 223, 224
SnakeViz, 226
version 2.0, divide-and-conquer
technique, 229–232
version 3.0
FastClosestPair Class,
233–236, 238–241
sorting functions, 233
visualize calling hierarchy, 228
Collect_mse inner function, 274
combine function, 435
Community common
neighbors, 392
Community resource
allocation, 392

Complex domain, 398–400, 415

Complicated domain, 400, 415

Concepts/techniques, machine
learning
collinearity, 278–280
data_generator.py
module, 259, 260
error terms, 271
evaluate_model function, 263
features *vs.* outputs, 261
gaussian random variable, 260
MSE, 269
normal distribution, 260
observer.py
module, 263–266
overfitting, 271, 273–276
real world process, 259
regularization, 285
residuals plot
demo_residuals
function, 281–283
evaluate_model
function, 285
fit method, 284
set_params() method, 284
runtime model, 270
scikit-learn framework,
262, 263
session.py module, 268
trivial training process, 270
underfitting, 276–278
warning module, 269
%connect_info magic
command, 135

Consumer-driven contracts
(CDCs), 108

Cross-validation (CV) score, 273

Cyclomatic complexity
ascending order, 89
built-in function, 89
execution flows, 89
global evaluation score, 91
sorting routine, 90
spyder, 89

Cynefin domains,
quadrants, 420, 421

Cynefin framework, 399, 402, 415

D

Dask array, 206, 413

Dask dataframe, 413

Dask delivers, 413

Dask framework, 413

Data engineers, 29

DataLoader class, 445

Data modification attack, 364

Data preparation, 30

Data preprocessing, financial model
features
AAPL price log, 294
data_preprocessing.py
Module, 300, 301
data_visualization.py
module, 301, 302
driver.py module, 300, 302
histograms, 298, 299
log returns, 300

Data preprocessing,
    financial model (*cont.*)
      normalization, 298
      outliers, 296
      returns, 294
      volatility, 296, 297
    IPython console, 288
    Pandas data frame, 289
    style parameter, 290
    time series
      AAPL closing levels, 292
      moving average, 291
      scaling, 292
    timestamp, 290
Data processing
    abstractions *vs.* latent
      features, 201
    augmented ball descend case
      study (*see* Augmented ball
      descend)
    compressing rating
      matrix, 202–204
Data science project
    Brown Cow model, 4
    case study (Cholera), 4–6, 8
    learning, 13
      domain knowledge, 14, 15
      domain-related terms, 15
      programming
      experience, 16, 17
    old *vs.* modern, 11
    phases, 3
Data security
    backup procedure, 347

    collection, 348
    disclosure, 348
    incident regarding e-mail, 342
    inference, 348
    Microsoft security development
      lifecycle, 346
    OWSP, 344
    phases, 347
    problems, 343
    suspicious e-mails, 343
    tool samples, 345
    unsecured connection, 343
Data streaming, 423
Data type, 32, 289
Data visualization
    closest pair case study (*see*
      Closest pair algorithm)
    matplotlib architecture,
      case study
      geographic map, 211–213
      higher-level components,
      210, 211
      plotting temperatures, 213,
      214, 216–218
Data wrangling, 69
Deep learning, 427–448
Degree centrality, 389
demo_metrics_and_mse
    function, 267
demo_regularization function, 286
Digit, counting, 402, 403, 406
Disorder, 399
Divide-and-conquer
    technique, 229

Divisible numbers, counting, 413

Document structure
    abstract, 150
    data science life cycle
        phases, 151
    dataset, 150
    drawbacks, 151
    future work, 151
    motivation, 150
    project (*see* Wikipedia edits
        project)
    references, 151

Domain-specific language
    (DSL), 244

driver.py script
    program, restructure, 68
    reusability, 68

E

E-Commerce customer
    segmentation
    CSV format, 30
    EDA, 30
    marketing domain, 32
    project in Spyder (*see* Spyder
        project)
    structured data, 31

Edge betweenness centrality,
    409, 411

Euclidian distance, 218, 222

European economic area
    (EEA), 349

expectedGain function, 448

Exploratory data analysis (EDA),
    30, 211, 288, 415, 419

extract_config method, 335

F

Feature engineering
    column types, 41
    custom
        Age_Group, 45, 46
        bar plots, 48, 49
        CTR, 46, 48, 51
        multilevel data frame, 53, 54
        NaN, 47
        nonzero method, 47
        number of clicks, 51, 52
    defined, 38
    describe method, 40
    dummy variables, 41
    e-commerce, 45
    logged-out category, 44, 45
    pandas package, 39
    scatter plots, 42, 43
    SciPy, 38
    summarize function, 54, 55

Filter bubble, 337

Financial modeling
    data preprocessing (*see* Data
        preprocessing, financial
        model)
    data retrieval, 288
    feature engineering
        correlation matrix, 307, 308
        create features, 306

Financial modeling (*cont.*)
  feature_engineering.py
    Module, 304
  heat map plotting, 307
  log returns, 305
  mean reversion, 303
  scatter plot, 304
  TA-Lib, 306
  target feature, 305
  time series analysis, 286
  timestamping, 287
find_path method, 188, 196
Fixing bug, software methods
  agile methods, 93
  correct version, code, 96, 97
    language of business, 98, 99
  defect code, 94
  improved fix, 97
  unit test, 94–96
  vectorized version, 100
    business-associated
      arguments, 101, 102
    classical arguments, 101
    NumPy framework, 100
Full-batch learning, 256
Funny elevator case study
  divide and conquer paradigm,
    115, 117
  socio-economic/socio-
    technical aspects, 117
  testing, 113
  unoptimized variant, 112

G

General data protection regulation
    (GDPR), 79, 341
  absolutistic approach, 353
  architecture, 357
  attribute-based access control
    model, 355
  behavioral patterns, 349
  controllers/processors,
    regulation, 353
  data breach, 354
  DevOps paradigm, 358
  domain-driven design, 358
  EU-based organizations, 349
  fundamental rights, 350
  health care system, 358
  lawful processing, 351
  microservices, 358
  remote access, 352
  requests, 352
  risk management, 354
  security measures, 353
generate_points method, 223
Geocoding API, 66
GitHub repository, 66
Global Historical Climatology
    Network-Daily (GHCN-
    Daily), 210
Graph analysis
  bipartite graphs, 377–379
  quality attributes, 376, 377

social networks, 369, 385–394
usage matrix
    built-in matplotlib
      engine, 376
    generic template, 371
    Graphviz, 375
    multigraph, 370
    NetworkX, 370
    optimization task, 370
    sample instance, 372
    UCs, 370
    UML use case diagram, 371
    usage_matrix.py Module,
      consent, 373, 374

H

High-dimensional datasets, 415
Homoscedastic outputs, 260

I

in_degree method, 376
Intelligent machines, 428–430
IPyWidgets, 145
i-th shadow model, 362

J

Jaccard coefficient, 392
JupyterHub, 124
JupyterLab, 124–126
    Anaconda Navigator, 125
    code execution
      abrupt stoppage, 130

    cells, 128
    doctest tests, 133
    error message, 128
    Hanoi Tower Solver, 127,
      130, 132
    HTML, 129
    notebook, 126, 128
    output, 133
    project (*see* Ball Descend
      project)
    screen, 125, 126
    simulator notebook, 148, 149
Jupyter Notebook, 124
Jupyter project
    notebook, 122
    principal components, 122, 123
    tools, 124, 125
Jupyter widgets and notebook
    extensions, 125

K

Kafka, 70
Kaggle, 66, 152
KDD Cup 1999 Data, 416
Kernel, 134–136
K-fold CV, 274

L

Local clustering coefficient
    (LCC), 388
Label flipping, 363
Label modification attack, 363

Large Hadron Collider (LHC), 12

Large-scale software system
adaptive maintenance, 77
corrective maintenance, 76
definition, 75
fixing bugs, 76
holistic approach, 76
knowledge areas, 79
life cycle model, 75, 77
preparation/planning phase, 75
preventive correction, 78
preventive maintenance, 77
scope, 80
tournament chess game,
analogy, 75
types, 78

Las Vegas model, 412

Lehman's laws of software
evolution, 78

LensKit for Python (LKPY)
extract_config method, 335
Fallback class, 330
knn wrapper package, 330
lkpy_demo.py file, 332
MultiEval facility, 333
nDCG Top-N accuracy metric,
335, 336
package structure, 330
PMML serialized models, 336
README.txt file, 331
SciPy style, 331
UML class diagram, 329
UserUser algorithm, 334, 335

Linear regression, 262

nonparametric, 262
parametric, 262
semiparametric, 262

%load lkpy_demo.py, 332

Logistic regression, 280

Log returns, 294, 296, 298, 299, 302, 304, 305

M

Machine-based feature
engineering
implementation from scratch,
439, 442, 443
implementation with PyTorch,
443–445

Machine learning
big data, 255
concepts/techniques (*see*
Concepts/techniques,
machine learning)
methods, 256–258
styles, 256

Machine learning as a service
(MLaaS), 360

Manual feature engineering
aggregated state, 433–435
combine function, 435, 436

Massive open online course
(MOOC), 10–12

Mean squared error (MSE),
263, 317

Membership inference attack
black-box, 360

MLaaS, 360
overfitted model, 361
training dataset, 361
white-box, 360
Mini-batch learning, 257
Miniconda distribution, 18
Monstrous models, 275

N

Naive algorithm, 406
Naming abstractions, 25
nbconvert, 125, 149
nbviewer, 125
NetworkX, 369, 370, 372, 377,
    379, 380, 382, 386, 393,
    394, 396, 409
next_neighbor method, 169,
    195, 196
Not a Number (NaN), 47
NumPy package, 140

O

Old *vs.* Modern data science
    projects, 11
Online learning, 257
Open-source software (OSS), 122,
    185, 344, 346
OpenStreetMap, 67
Open triads, 389
Open Web Application Security
    Project (OWASP), 344

P

pairplot function, 416
Path finder
    find_path function, 144
    terrain, 142, 143
    top-down decomposition,
        140, 142
    wall function, 140, 141
Perl module, 68
plot_mse function, 263, 272, 312
Poisoning attack
    categories, 363
    linear regression model, 364
Poison insertion attack, 363
Policy authoring point (PAP), 357
Policy decision point (PDP), 357
Policy enforcement point
    (PEP), 357
Policy information point (PIP), 357
Prediction accuracy metrics, 337
Predictive Model Markup
    Language (PMML), 315, 336
predict_proba methods, 401
Preferential attachment, 392
Principal component analysis
    (PCA) method, 15, 415
Privacy, meaning, 348, 349
Problem-solving ability, 398
Public data sources
    data analysis, 64
    Internet, accessible, 66, 67
%pwd magic command, 37
    scikit-learn, 64, 65

Q

Quandl, 67
Quantitative data, 31

R

read_csv function, 39
Recommender systems
  machine learning, usage, 317
  mashup movie, 322
    JSON response, 322
    OMDb service, 325–327
    simple_movie_
      recommender.py Module,
      327, 328
    tastedrive_service.py
      Module, 324, 325
  MovieLens
    categories, 319
    collaborative filtering, 322
    content-based, 321
    context-based systems, 322
    information retrieval
      system, 320
    interfaces, 321
    relational database
      system, 320
    types, 318, 319
    untracked option, 320
  standard root mean squared
    error, 317
Reinforcement learning, 258
requests package, 35

Resilient distributed dataset
  (RDD), 196, 309
Ridge regression, 285, 286
runtime model, 259, 270

S

Scalable graph loading
  CSV file, 380
  load graph, 383–385
  nodes.csv file, content, 382
  standard edge list format, 382
Scaled percent returns, 294
Scatter matrix plot, 280, 281
SciPy ecosystem, 38, 206
Semi-supervised learning, 256
Shadow training
  attack model, 362
  highest-scored label, 362
  overall structure, 363
  target system, 362
  test sets, 362
Simple domain, 398, 399
Simple moving average
  (SMA), 291
sklearn.feature_selection
  module, 416
Social network
  centrality, 389
  eccentricity, 389
  karate club network, 390–393
  LCC, 388
  local/global graph metrics,
    386, 387

machine learning systems, 385

NetworkX's documentation, 386

prediction measures, 392, 393

single metric graph, 388

six degrees of separation, 394

Socio-* pieces of software
    production
  socio-economic aspects, 112
  socio-technical aspects, 112

Software engineering, 73
  agile principles, 83
  application, 74
  bug free legacy code, 103, 104
  context awareness and
    communication
    bug (*see* Fixing bug, software
     engineering)
    cone of uncertainty, 91–93
    constant feedback
     loop, 85, 88
    context and knowledge
     oblivious, 85, 86
    cyclomatic complexity,
     89–91
    knowledgeable, 85, 87
  conventions, 81
  faculty code, 105, 107
  large-scale software systems, 74
  legacy code, 102
  Python ecosystem, 83, 84
  quality assurance tools, 83
  risk, 75
  rules and principles, 81, 82

Software engineering *vs.* data
    science, 1

Source lines of code (SLOC), 84

Spyder project
  chunk data, 62
  code restructure, CSV file,
    62–64
  dataset
    download, 34
    feature, association
     (*see* Feature engineering)
    file sizes, 38
    IPython console, 19, 36
    shell-related magic
     commands, 37
  driver code, 36
  folder structure, 34
  nyt_data.py Script, 34, 35
  results
    Apache Parquet, 60
    driver.py, 60
    multilevel index, 61
    Parquet engine, 61
  summarize function, 62
  test pipeline
    click-through rate, 57
    get_file_number function, 56
    number of clicks, 59
    traverse function, 56, 57
  UI, 19

Staged processing, 258

Stochastic gradient descent,
    257, 297

Streaming linear regression
    Apache Spark's MLlib
        framework, 308
    driver.py Module, 311
    RDD, 309
    setIntercept method, 312
    streaming_regression.py
        Module, 309–311
    training and test streams,
        313, 314
StreamingLinearRegression
    WithSGD class, 308
Streaming system, 257
Stream processing, 287
Structured data, 31
style parameter, 290
Sum of squared errors (SSE), 269
Supervised learning, 256, 275
Systems/software development life
    cycle (SDLC), 1

T

TA-Lib, 306
Total cost of ownership (TCO), 79
Transitivity, 389
t-SNE method
    features and components, 419
    power, 417
    variables, 417
Turing test setup, 430

U

UCI Machine Learning
    Repository, 67
Unbiased minimal variance
    estimator, 269
Unsupervised learning, 256
Use cases (UCs), 370

V

VarianceThreshold class, 416
Variety, 9
Velocity, 10
Veracity, 10
Visualization, 38, 209
Volume, 9

W, X, Y

Wikipedia edits project
    abstract, 152
    data, 154
    drawbacks, 153
    JupyterLab notebook, 156
    motivation, 152
    specification fixing, 155

Z

zipfile package, 35

CPSIA information can be obtained
at www.ICGtesting.com
Printed in the USA
LVHW052100140120
643594LV00006B/105